D0237037

WHAT THE
BRITISH
INVENTED

WHAT THE BRITISH INVENTED

FROM THE GREAT TO THE DOWNRIGHT BONKERS

GILLY PICKUP

AMBERLEY

Than ... *skills,*
editor ... *coffee.*

First published 2015

Amberley Publishing
The Hill, Stroud
Gloucestershire, GL5 4EP

www.amberleybooks.com

British Library Cataloguing in Publication Data.
A catalogue record for this book is available from the British Library.

ISBN 978 1 4456 5027 2 (print)
ISBN 978 1 4456 5028 9 (ebook)

Typesetting and Origination by Amberley Publishing.
Printed in Great Britain.

Contents

	Introduction	7
1	Domestic	9
2	Movement	59
3	Technology	105
4	Medicine	137
5	Sport	161
6	Miscellaneous	189
7	Modern-Day Inventions	230
	Bibliography & Sources	243
	Index of Inventions	250

Introduction

The world would be a much poorer place without our Great British Inventions – from catseyes to crossword puzzles, tarmac to telephones, steam engines to shorthand and pneumatic tyres to penicillin, British inventions have led the world.

We Brits are a creative lot, also responsible for lawnmowers, radar, fire extinguishers, tin cans, chocolate bars, hypnotism and even the sandwich. Perhaps, however, the idea which has most dramatically changed all our lives is the World Wide Web – not the internet – developed by Londoner Tim Berners-Lee in 1989.

Believe it or not, the Bank of France was also the brainchild of a Brit, a Scot to be precise – John Law, who was appointed Controller General of Finances of France under King Louis XV. Then there was the entrepreneurial Hubert Cecil Booth, who invented the 'Puffing Billy', the first powered vacuum cleaner, while John Walker was the bright spark who discovered matches in 1827. He coated the end of a piece of stick with chemicals that burst into flames when rubbed against a rough surface. And where would we be without flush loos? Invented by Sir John Harrington, not Thomas Crapper as many maintain, although Crapper was a nineteenth-century plumber who patented a few bathroom fittings …

These ideas are dreamed up in eureka moments, when chance and inspiration combine to create something wonderful. So, without further ado, let us take a closer look at those brilliant, sometimes slightly bonkers British men

– yes, ladies, sadly for us they are almost all male – who have not only done so much to improve our daily lives, but who have also changed the world around us.

Invent: To produce or contrive (something previously unknown) by the use of ingenuity or imagination.

1

Domestic

Flush Toilet

I could say that after he invented it he was probably flushed with success, but that is probably an old joke. Anyway, back in the 1590s, poet **Sir John Harrington** (sometimes spelled 'Harington') was banished from court for telling the ladies risqué stories and rude jokes. He was Queen Elizabeth I's godson, educated at Eton College and Cambridge University. During his exile at Kelston, near Bath, he built a house and devised the first flushing lavatory, which he named Ajax – 'jakes' was a slang word of the day for 'toilet'.

In those days people just used buckets, although the wealthy would sometimes have a close stool, which would have a velvet padded seat with a container beneath it. Of course, some unfortunate soul still had to empty it into a river or moat. This made Harrington's invention all the more unusual, as his Ajax emptied itself. When a handle on the seat was turned, water carried the contents into a cesspool below. He proudly proclaimed that it 'would make unsavoury places sweet, noisome places wholesome and filthy places cleanly'.

In time the queen forgave him and visited his house. His invention impressed her, and she tried it out herself. Harrington's water closet had a pan with an opening at the bottom that was sealed with a leather-faced valve. Handles, levers and weights poured in water from a cistern, opening the valve. Harrington had one installed for Her Majesty at

William III's close stool.
(© Historic Royal Palaces)

Richmond Palace. Whether or not she was enthusiastic about the newfangled device is not recorded, but then she didn't have to empty her own close stool anyway.

Harrington is also remembered for the political epigram, 'Treason doth never prosper: what's the reason? Why, if it prosper, none dare call it treason.'

But There's More …
Many people believe that nineteenth-century plumber **Thomas Crapper** from Yorkshire was the inventor of the lavatory. He was a clever chap who promoted bathroom fittings and bought the rights to a patent for a 'Silent Valveless Water Waste Preventer'. He cannot be given credit for the flush toilet, though he did make some improvements to the flushing system.

In 1863 he invented a contraption for use with WCs, the self-rising seat, but the idea was unsuccessful. However, his standing was increased in the 1880s when Queen Victoria bought Sandringham House. When the old house was demolished to make way for a new one, Crapper & Co were asked to supply the bathroom wares and plumbing, gaining him a royal warrant. Manhole covers in Westminster Abbey with Crapper's company name on them are now one of London's minor tourist attractions.

An interesting endnote: Thomas grew wealthy, and he and his brother George would often go to the Finborough Arms in Kensington to start their working day with a bottle of champagne!

Then there is the S-bend, another important part of the modern flush toilet still in use, which was invented by Edinburgh-born **Alexander Cummings** in 1775. He received the world's first patent for a flushing toilet. His idea used standing water in the toilet bowl to prevent horrible smells from backing up out of the sewer. His premises were in London's Bond Street. Cummings was a master of many arts, and as well being an inventor he was a mathematician, mechanic and watchmaker.

Joseph Bramah, a Yorkshire locksmith, patented the first practical water closet in England in 1778. He worked to improve on Cummings' design as he discovered that the model then being installed in London houses was apt to freeze in cold weather.

He used a plunger to release the waste and seal the soil pipe as well, and he invented a float valve system for the cistern. Today, a Bramah is still in use at the House of Lords.

Prolific inventor Brama-h, best known for having invented the hydraulic press in 1795 (two piston cylinders with different cross-sectional areas connected with a tube and filled with fluid so that moving one piston causes the other to move), obtained eighteen patents for his designs between 1778 and 1812. These included a beer engine (1797), a planing machine (1802), a paper-making machine (1805), a machine for automatically printing banknotes with sequential serial numbers (1806) and the fountain pen (1809). He also patented the first extrusion process for making lead pipes, and machinery for making gun stocks. That sounds complicated. Did he ever stop inventing?

And Still More ...

Thomas Twyford from Staffordshire was a pottery manufacturer who invented the single-piece, valveless ceramic toilet in 1883. This was much easier to clean than previous wood or metal models. It was hygienic and free-standing and the whole appliance was fully exposed. Advertisements of the day stated, 'No filth, nor anything causing offensive smells can accumulate or escape detection.' It was described as a 'Perfection of Cleanliness'. Vast quantities were manufactured and exported and installations were made in Buckingham Palace.

World's First Public Toilets

Before the 1850s, when you were out and about it was to be hoped that you didn't feel the urge to 'spend a penny' as public toilets hadn't yet been imagined. So it's three cheers for **Josiah George Jennings**, a plumber from Hampshire who brought relief to many when he invented just that. They were installed at the 1851 Great Exhibition in Crystal Palace and generated much excitement. They were called 'Monkey Closets' and were installed in the Retiring Rooms there. Records show that more than 827,000 visitors paid one penny to use them – and yes, that's where we get the phrase 'spend a penny'. It was a pretty good deal, really, because for your penny you also got a clean seat, a comb, a towel and even a shoeshine.

But It Would All Be Useless Without a Sewage System
In 1865 **Joseph Bazalgette** of London, chief engineer of London's Metropolitan Board of Works, developed the biggest intersecting sewer system the world had ever seen.

In the mid nineteenth century, London was having a terrible problem with recurring epidemics of cholera. In 1853–4 more than 10,000 Londoners died from the disease, which was then thought to be caused by foul air. Dr John Snow said this was nonsense and that the outbreaks were waterborne, but no one believed him. Thank goodness he eventually proved them wrong after he traced an outbreak to a Soho water pump.

Anyway, back to Bazalgette, probably best described as an engineer rather than an inventor, who created the sewer system in response to the 'Great Stink' of 1858. It was created by the hot summer that year and was caused by raw sewage being dumped in the Thames ... what's even more gruesome, though, is that the river still continued to supply the city's drinking water! The dreadful stench affected all those who went near the Thames, which of course also meant those in Westminster, who tried to combat the smell by applying chemicals in their meeting rooms.

This gave impetus to legislation enabling the Metropolitan Board to begin work on sewers, and, thanks to Bazalgette, most of London was connected to a sewer network by 1866. The flow of foul water from old sewers and underground rivers was intercepted and diverted along new, low-level sewers built behind embankments on the riverfront and then pumped eastwards out to sea.

This system, so instrumental in relieving the city of cholera epidemics and cleansing the river, is still in use today.

In days of yore, chamber pots were usually emptied from an upstairs window into the street below, regardless of anyone who happened to be passing at the time. It meant you had to be jolly careful where you walked. The cry 'Gardy loo!' came from the French *gardez-l'eau*, which meant 'watch out for the water'. This was the origin of the British nickname for the lavatory, the 'loo'.

Toothbrush

Incarcerated in London's Newgate Prison in the 1770s for provoking a riot in Spitalfields, forty-six-year-old businessman **William Addis** probably felt a little bored. He was a stationer and rag merchant who supplied finished paper to the book trade. The rags he harvested were pulped and remade into sheets of paper so nothing went to waste. William's clients, London booksellers of the eighteenth century, also sold patent medicines and supplies for pharmacists. It seems curious today, but it is no stranger than the barber surgeons of yesteryear, who did everything from cutting hair to pulling teeth and minor operations.

One morning in his cell, William began to clean his teeth. He cleaned them much in the way that most people did in those days, rubbing them with a cloth covered with soot or salt. Going back to days of yore, this had always been accepted practice. Aristotle had advised Alexander the Great to use a rag on his teeth, while George Washington's dentist suggested using a rag with some chalk on it.

Using the same method, and with little else to think about, it dawned on Addis that this was actually not terribly effective. It was fine up to a point, but meant it was impossible to get in-between the teeth to clean them properly. By the following day, an idea had planted itself in his head. He saved a small bone from the meat he was served, bored small holes into it, then acquired some hard bristles through his helpful prison guard. He cut them and tied them in tufts, using glue to seal the tied end of the bristles to the bone. So there it was, the world's first toothbrush.

Of course, success is often due to timing or being in the right place at the right time. William's toothbrush idea came at exactly the right time. Refined sugar, unknown in London in medieval times, was now being consumed in industrial quantities as supplies came from the West Indies. No wonder Georgian Londoners had awful teeth, but what could they do? After all, effective dental repairs were still at least a century away. The only option was extraction, a painful, barbaric process carried out by those barber surgeons.

Released from gaol, William was raring to go. He knew that his new idea would be a winner. He set to and produced a small number of toothbrushes made from animal bone and horsehair. He offered them for sale to his book trade contacts. They were instantly popular – anything that might reduce the chances of visiting a barber surgeon just had to be a good thing. The toothbrush became a must-have item in Georgian London. It wasn't long before other manufacturers copied William's idea, but fortunately it didn't stop the company's expansion; nor did it prevent William growing rich.

By 1840 his son, who was also called William, ran the company, which by then had sixty employees. They used a system which was widespread in London's East End in those days: piecework, with the women of Spitalfields and Whitechapel producing brushes in their own homes.

The women were paid according to how much they produced, though they had to buy their own materials up front. It was tough, because if the goods did not come up to scratch then the women didn't get paid. Well, those were pre-union days after all. In the meantime, the Addises grew very wealthy indeed. By then they were producing four different models of toothbrush: gents', ladies', child's and Tom Thumb. William used badger hair for the most expensive brushes, with hog, pig or boar hair, all imported, for the cheaper versions. This mainly came from Russia, Poland, Bulgaria and Romania. Nowadays, of course, we would turn our noses up at the thought of cleaning our teeth with bone and animal hair, but it wasn't until 1935 that an American discovered nylon. The Addis company, which was by now Wisdom Toothbrushes, was quick to jump on the bandwagon and, agreeing to a deal with UK licensee ICI, produced the first synthetic toothbrushes.

The new brush was more expensive than the bone-and-hair competition at two shillings, but again it was all about timing. It was 1940, and British housewives were being told to waste nothing, using bones to make stock in soups and stews. This shortage of bone worked in the Addises' favour.

When the last member of the Addis family left the firm in 1996, it brought to an end over two hundred years of history. William Addis's invention was one that changed the world. Just think, from bones and badger hair to toothbrushes. Where would we be without them?

Tooth-cleaning methods have been around since the dawn of civilisation. Methods employed in ancient communities in India, Arabia and China included the use of plant stalks, twigs from trees, feathers and quills or chalk.

Chocolate

It was fantastic when eighteenth-century France produced tablets and bars, but it wasn't until 1847 that **Fry & Son** made the first bar of chocolate as we know it today. It was blended from a mixture of cocoa powder and sugar with a little melted cocoa butter that had been extracted from the beans. It tasted coarse and bitter by today's standards, but nevertheless it was still a revolution. It was sold to the public as chocolate *delicieux a manger* – delicious to eat – because, until this point, chocolate had only been consumed as a drink. So, chocoholics everywhere have cause to thank Fry & Son.

Not that the Fry family were only responsible for bringing the chocolate bar to us. They were pretty nifty with other inventions too. **John Doyle Fry** and his brother-in-law **Robert Barclay** were printers who invented and patented the chequebook and, later, a method of printing on metal which

An 1851 chocolate bar. (© Frenchay Village Museum)

produced colourful tins for chocolate. The firm lives on today as the Metal Box Company.

Jeremy Fry, son of the last Fry to head the chocolate firm, became a product designer with Frenchay Products Ltd between 1954 and 1957. He founded Rotork Engineering Company in 1957 after identifying the potential of valve actuators (mechanised devices that automatically move a valve to a desired position) and sold his inventions through his company. Initially Rotork had no more than a dozen employees, but the company was soon taking advantage of the expansion of the oil industry in Europe. Rotork valve actuators are used across the world in areas ranging from power plants to marine engineering.

Jeremy Fry nurtured the talents of his employees. In the late 1960s he spotted the young James Dyson, recently graduated from the Royal College of Art, and collaborated with him on the construction of the theatre auditorium inside the Roundhouse.

With Dyson he invented the sea truck, a high-speed maritime vehicle that skims along the water on a layer of bubbles. Dyson said recently of Fry that 'he ... gave enormous responsibility to young people who had no real experience'.

In 1873, Fry, Vaughan & Co. produced the first chocolate Easter egg in Bristol.

Worcestershire Sauce

The idea for Worcestershire sauce came from a wealthy man who lived for a time in India and wanted someone to recreate the recipe he brought home with him. He asked dispensing chemists **John Wheeley Lea** and **William Henry Perrins** to produce it for him, which they did at their premises at 60 Broad Street, Worcester.

Worcestershire sauce, sometimes shortened to Worcester sauce, is a fermented liquid condiment primarily used to flavour meat or fish dishes. When the recipe was first mixed at the pharmacy, the result was so strong that the makers thought it was useless. In fact, it tasted so awful that they thought there was nothing else for it but to abandon the barrel in the basement. When space in the storage area was required some years later, the chemists decided to try it again and to their amazement discovered that the sauce had fermented and mellowed and was now palatable.

The Lea & Perrins brand was commercialised in 1837 and has been produced in the current Midlands Road factory in Worcester since 16 October 1897. In 1838 the first bottles of 'Lea & Perrins Worcestershire sauce' were released to the general public, but it was sold originally as much as a medicine as a condiment.

Lea & Perrins advertisement, 1900.

In 1930, the Lea & Perrins operation was purchased by HP Foods, which was in turn acquired by the Imperial Tobacco Company in 1967. HP was sold to Danone in 1988 and then to Heinz in 2005.

The ingredients of a traditional bottle of Worcestershire sauce sold in the UK – to give it its proper name the Original & Genuine Lea & Perrins Worcestershire Sauce – are malt vinegar, spirit vinegar, molasses, sugar, salt, anchovies, tamarind extract, onions, garlic, spice and flavouring. The 'spice and flavouring' is believed to include cloves, soy sauce, lemons, pickles and peppers.

Seaside Rock

Invented around 1887 by **Ben Bullock** from Lancashire, seaside rock was initially made for factory workers to take home as a gift after a day at the seaside. Ben, originally a miner who went on to run his own sweet factory in Dewsbury, produced the first seaside rock, which was pink on the outside and had letters running all the way through. Pleased as punch with his idea – after all, it was rather a good one – he sent batches of it to shops in Blackpool and before long it began to be produced for seaside resorts all over England. He was on to a winner.

Before seaside rock made its appearance, there was a similarly shaped treat called fair rock, so called because it was sold at fairgrounds in Britain. The difference between the two was that fair rock was uncoloured and had no letters running through the centre.

So, how is seaside rock made? A craftsman called a sugar boiler is responsible. Ingredients are sugar and glucose, with lots more sugar than glucose, mixed with water and boiled to

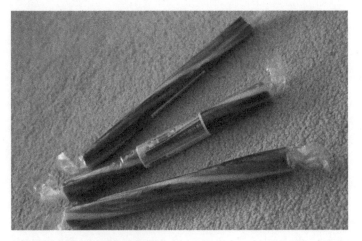

Seaside rock. (© Gilly Pickup)

almost 150 degrees Centigrade. After cooling it is separated into parts. The job of the sugar boiler's assistant is to take the larger, uncoloured portion of the batch to 'pull'. Pulling used to be done manually but nowadays a machine is used. The movement of the machine helps bring air into the mass, turning it from golden into the white, cloudy appearance that forms the centre of the stick of rock. The outer layer and lettering sections are coloured.

Then there are the letters, which are made individually before being stuck together in a line with white filler. It is certainly a palaver to get the lettering correct. It is not surprising that it can take up to ten years to master, as rock is often up to 6 feet long before it is cut. Square-shaped letters – B, E, F, K – and triangular letters – A and V – are made first, while round letters – C, D, O and Q – are made last to prevent loss of shape before the rock sets. The lettering, filling and core are rolled together then wrapped in the brightly coloured outer casing. The whole slab is then stretched into smaller, longer strips by machine before being cut and wrapped ready for sale.

While it is generally agreed that Ben was the brains behind producing this dentistry nightmare, a character called 'Dynamite Dick' from Blackpool is occasionally given the credit for being first to add lettering to the rock.

Ben's first batch of lettered rock had the words 'Whoa Emma' running through it as a tribute to a popular song of the day.

And while on the subject of sweets, did you know that Liquorice Allsorts are British too? The colourful sweets were made by accident. In 1899, a sales rep named **Charlie Thompson** who worked for the Bassett Company went to visit a wholesaler. He was a tad nervous about trying to sell his wares and accidentally – and fortuitously, as it happened – dropped a whole tray of different types of sweet samples on the floor. Hugely embarrassed, Thompson got on his hands and knees to pick them up, but the now interested customer asked if he could order them all as a mixture. Liquorice allsorts, also spelled licorice allsorts, were born. Owing to the success of liquorice products such as Ju-Jubes and Liquorice Allsorts, Bassett's created other brands including Victory V Lozenges, Zubes, Sherbet Fountains and Beech Nut. How many of you remember them?

Mascot Bertie Bassett was created in 1926 and still continues to represent the brand. A figure made up of liquorice allsorts, he has become a part of British popular culture.

Did you know liquorice is good for you? History tells that, when they were on the march, Alexander the Great's troops were supplied with extract of the liquorice plant root to quench their thirst.

Marmalade

According to one story, the name 'marmalade' originates from *Marie malade*, which translates as 'Mary's illness', a reference to Mary, Queen of Scots, whose doctor prescribed oranges mixed with sugar for her seasickness.

Another story tells that marmalade itself was invented in 1700. A storm-damaged Spanish ship carrying Seville oranges sought refuge in Dundee harbour. James Keiller was a poor local merchant who managed to buy the oranges for a pittance, and his wife Janet then turned them into the preserve we know today. As time passed, embellishments were added to the story, one being that Janet's little boy carried loads of the fruit from the ship to the beach. His mum sent him back to get even more by telling him, 'Mair, ma lad!' In English, this translates as 'More, my lad!'

Perhaps that's just a nice story, though, because in a recipe book dated around 1677 belonging to Eliza Cholmondeley from Chester there is a recipe for a Marmelet of Oranges which is similar to the marmalade we know today.

In the early 1700s, cooks used a preserve akin to marmalade as an ingredient and accompaniment. Robert Smith, who was a cook to King William III and the French ambassador, the Duc d'Aumont, published a cookery book in 1725. His *Court Cookery, or the Compleat English Cook* gives a recipe for rice pudding made with a layer of orange marmalade at the bottom as well as pieces of beef marrow to add richness.

Of course there are others out there who could claim that marmalade was their idea. The Romans ate a fruit preserve made from quinces and honey which was known as *marmelo*. They ate it to keep evil spirits away. Then there is the Greek *marmelada* and the Italian *marmelatta*.

What is without doubt, though, is that the Keillers of Dundee were the first to establish a marmalade factory.

Sorting peel for marmalade. (© Lester Moyse)

James and Janet were instrumental in the popularisation of marmalade and were probably the first commercial producers of the preserve. For the first half of the nineteenth century the affordable Keillers Dundee Marmalade was enormously popular and the forerunner of today's best-selling brands ...

... And of course we all know that Paddington Bear loves marmalade and would put nothing else in his sandwiches.

One of Janet Keiller's great-great-great-great-grandsons is television presenter Monty Don.

Speaking of marmalade brings to mind ...

The Toaster

That most essential of kitchen goods, the toaster was invented by Scot **Alan MacMasters** in 1893. Macmasters' toaster was commercialised by Crompton & Co. in the same year. They called it the Eclipse. There was a problem with it, though, in that it was difficult to obtain a metal that was suitable. Iron wiring melted easily, and of course presented a fire hazard. An alloy of nickel and chromium was patented that allowed current to be passed and a dissipation of heat to occur without the element melting. It was refined in 1918 to allow doors to turn the toast over, allowing toasting on both sides. Great progress when you think of how the only way to toast bread previous to that was by putting it in a metal frame or sticking a slice of bread on the end of a toasting fork and holding it in front of an open fire. So was Alan's toaster the best thing since sliced bread? Maybe. We can't take credit for that, though, because it was an American, Otto Frederick Rohwedder, not a Brit, who invented sliced bread in 1928.

Now, it's all very well to have toast and marmalade, but we Brits need our morning cuppa to accompany it ...

The Automatic Kettle

Another British invention. **Peter Hobbs** was the managing director of a division of Morphy Richards who, in 1951, met an employee called Bill Russell. When Hobbs had a disagreement with Charles Richards, he decided to leave the company. He wanted to design an electric coffee percolator and discussed it with Bill Russell, and they set up the company called Russell Hobbs in Croydon, Surrey. The first coffee percolator in the world appeared in 1952.

The automatic kettle, one that switches itself off when the water boils, appeared in 1955. It was a fairly simple piece of technology, focussing on a bimetallic strip which bent as the water boiled, breaking a circuit and switching off the kettle. Prior to the automatic kettle, kettles had a habit of boiling dry – or worse, starting a fire – if left unattended.

Over the years Russell Hobbs continued to lead the way in kettle manufacturing, and in 1960 the iconic K2 kettle

– chrome and curved, with a red button sticking out of its plastic handle – was born. This became the must-have kitchen appliance during the 1960s and 1970s and was the best-selling kettle in Britain. It cost about £7, when the weekly average wage was £14. Many brides-to-be included a K2 on their wedding wish list.

Russell Hobbs has reports of K2s still going strong after thirty years, and these days collectors will pay a few hundred pounds for a K2 in mint condition.

Teasmade

Fancy a brew? Of course, some of you who are reading this might not know exactly what a 'Teasmade' is, so let me explain that it is a device which sits by your bed and wakes you up in the morning with a hot cup of tea. Great idea, but for some reason this machine, which has been on the go in one form or another for over a hundred years, never really took off.

Its history goes back to 17 December 1891, when **Samuel Rowbottom** of Derby applied for a patent for his Automatic Tea Making Apparatus, the patent being granted in 1892. It sounds as if it was a pretty dangerous device, consisting of a clockwork alarm clock, gas ring and pilot light. Although there is no evidence that he commercially produced his tea maker, the concept he invented of using the steam from boiling water to force the water out through a tube into the teapot is still in use today.

Further down the line, on 7 April 1902, a patent for a Teasmade was registered by gunsmith **Frank Clarke** of Birmingham. It was made of copper and brass and had an alarm clock which triggered a switch which in turn lit a match by striking it against moving sandpaper. This was then supposed to light the spirit stove under the kettle. When the kettle boiled, the pressure from the steam lifted a flap and made the kettle tilt over to fill the waiting teapot. Seems a lot of palaver to get a cup of tea. It didn't have a catchy name; it was given the title An Apparatus Whereby a Cup of Tea or Coffee is Automatically Made. That wasn't going to entice anyone to buy it, so it was later marketed as A Clock That Makes Tea. Not much better, really.

Then, on 2 May 1932, **George Absolom** from Sussex submitted an application for a patent on his invention, an electric, automatic tea maker. The patent, number 400672, was passed in November 1933. This invention was manufactured and marketed as the Teesmade, but although Absolom applied for a registered design using Teesmade as the name the Patent Office would not accept it. It was declined on the grounds that the unit was not made on the River Tees and the public might be confused. Geographic trademarks were usually refused at this time. However, although the name could not be formally protected, from 1932 onwards George Absolom traded as Teesmade Co. The first tea maker to be sold under the name Teasmade was invented by **William Hermann Brenner Thornton** and made by Goblin in 1934.

Having been made by a number of different companies across its history, the Teasmade ceased production in 2000. Shame, really.

The Sandwich

There is a well-founded claim that **John Montagu**, 4th Earl of Sandwich, invented the sandwich as we know it. A rumour in a contemporary travel book called *Tour to London* by Pierre

Sandwiches. (© Food4Media)

Jean Grosley formed the popular myth that bread and meat were what kept his Lordship on top form at the gambling table. As First Lord of the Admiralty, Montagu commanded the British Navy and was a noted explorer. One night in 1762 the earl was doing his favourite thing – gambling. Although he was peckish, he didn't want to stop for something to eat so told a waiter to bring him some beef placed between two chunks of bread. This meant he didn't have to stop his game to eat and it meant he didn't get the playing cards, or his fingers, covered with grease. His gambling cronies thought it was a good idea, and before long others began to request an order: 'The same as Sandwich!' At that time, though, sandwiches were mainly eaten by men at late-night drinking parties and it wasn't until the end of the eighteenth century that it gained popularity as a supper food. During the late nineteenth and early twentieth centuries many taverns offered free sandwiches to drinkers in order to attract customers.

And that is the story of how gambling led to the sandwich.

> The longest sandwich, created in Italy in 2004, was 2,081 feet. The largest sandwich ever made weighed 5,440 lbs.

The Tradition of Afternoon Tea

Anna, 7th Duchess of Bedford, is said to have introduced afternoon tea to Britain in 1840. It was fashionable in her upper-class circles to eat evening meals around 8 p.m., which meant there was a long gap between lunch and dinner. Anna didn't want to wait that long before eating and asked her maidservant to bring her tea, bread, butter and cake to stave off hunger pangs during the late afternoon. Soon she began inviting her friends to join her. The habit became a chic social event and ladies dressed in long gowns, gloves and hats would take afternoon tea in their drawing rooms between four and five o'clock in the afternoon. Tables, elaborately set with lace linen, would display a host of accessories including tea knives, cake stands, sandwich trays, tea caddies, sugar tongs, blending bowls and tea strainers.

However, it wasn't until 1864 that London bakery the Aerated Bread Company opened the first ever tearoom in Britain. Although tea gardens had been popular in the eighteenth century, these were only open during summer. They charged for admission and refused entry to the working classes. Joe Lyons soon became the biggest name in the business. He opened his first tea shop on London's Piccadilly in 1894, and the first of his famous corner houses fifteen years later.

Afternoon tea. (© Adrian Houseon/Oxford Randolph Hotel)

Nowadays traditional afternoon tea still consists of dainty crustless sandwiches, perhaps filled with cucumber, watercress or smoked salmon, scones served with clotted cream and preserves and at least one large cake, perhaps fruit, caraway seed or Madeira. English Breakfast, Indian or Ceylon tea, preferably loose leaf, should be poured from silver teapots into bone china cups.

A delicious accompaniment to afternoon tea is a Maids of Honour tart, a traditional English puff-pastry cake with sweet curd. Although there are various stories about the origin of the cakes, one story goes that King Henry VIII named them when he discovered Anne Boleyn and other maids of honour eating them in Richmond Palace. Apparently the bossy king was so delighted with the cakes that the recipe was kept secret and locked away. A tearoom in Kew called The Original Maids of Honour still serves the cakes, but the recipe – sssh! – remains a closely guarded secret.

The Original Maids of Honour Tearoom. (© Gilly Pickup)

Scotch Eggs

Well, it's all down to the name, isn't it? You could be forgiven for thinking that this one-time mainstay of picnics and lunch boxes originated north of the border, but that is not the case.

The story goes that the Scotch egg was invented by Fortnum & Mason, that exclusive London department store, back in the 1730s. At the time, Fortnum's were busy creating snack-type foods to suit their well-heeled customers, who tended to travel on long-distance carriage rides to their country houses.

Fortnum's started to produce travel-handy dishes like pork pies, which were wrapped up and put in baskets with cutlery. The Scotch egg was one of those foods.

For those readers who don't know exactly what this delightful titbit is, it is a hard-boiled egg wrapped in sausage meat, rolled in breadcrumbs and fried or baked. Substantial

Fortnum & Mason, 1857. (© Fortnum & Mason)

and tasty as well as easy to hold, it was – in fact, still is – an excellent way to stave off hunger pangs. The word 'Scotch', by the way, has nothing to do with Scotland; 'scotched' was merely another word for 'processed'.

Of course, like many claims, Fortnum's isn't the only theory about the origin of the Scotch egg and it may be the case that the company popularised an older product. It may be the case that Scotch eggs were an export from the British Raj like curry and kedgeree. According to the book *A Caledonian Feast* by Annette Hope, Scotch eggs could have been inspired by the Moghul dish *nargisi kofta* (narcissus meatballs). These were hard-boiled eggs coated with cooked spiced minced mutton and fried, then cut in half and served in a sauce of curried tomato and onion.

The earliest printed recipe for Scotch eggs appears to be in the 1809 edition of Mrs Rundell's *A New System of Domestic Cookery*. In those days, instead of sausage meat the meat would have been forcemeat, which was leftover bits of meat and offal mixed together to make mince. Back then Scotch eggs were often eaten hot, with gravy.

Of course, like many fads and fashions, the Scotch egg went out of favour; in fact, it became one of the most 'uncool' foods around. However, trends change and *Olive*, the food and cookery magazine, included Scotch eggs in its 2011 'cool list'. Gourmet versions now pop up in the most exclusive restaurants.

So there you are, a little modern-day menu inspiration for indoor picnics, instant lunches or, indeed, for long carriage rides.

> Fortnum's say they like to reinvent the Scotch egg from time to time, using duck, hen, quail, goose and ostrich eggs at the centre. They have mixed haggis with the sausage meat to make a more truly Scottish Scotch egg and even threw caution to the winds and replaced the sausage meat with salmon, coriander and sweet chilli, wrapped around a quail's egg. And, of course, they only use free-range eggs.

Arbroath Smokies

Now this is a delicacy which comes from north of the border, specifically from the village of Auchmithie, which is just outside Arbroath in Angus.

The story goes that this tasty dish originated as the result of a fire in a store which housed barrels of haddock preserved in salt. Next morning, when locals went to clear up the mess, they discovered the charred barrels of fish. Waste not, want not is a Scottish trait, and on trying the now cooked fish it was found to be extremely tasty.

Towards the end of the nineteenth century, when Arbroath's fishing industry went into decline, the town council offered Auchmithie's fisherfolk land in an area of the town known as the 'fit o' the toon' and the use of the harbour to boost the industry. Most of the population of Auchmithie relocated and their Arbroath smokie recipe came with them. A new international industry was born.

Today, several local businesses produce Arbroath smokies.

In 2004, the European Commission registered the designation 'Arbroath smokies' as a Protected Geographical Indication under the EU's Protected Food Name Scheme, acknowledging its unique status as a world-renowned product of quality.

Black Bun

This is a type of pastry-covered fruitcake that was originally eaten on Twelfth Night (5 January). Nowadays it is a traditional food eaten at Hogmanay and it is also given as a gift at first-footing ceremonies. It contains citrus peel, dried fruit, almonds, allspice, ginger, cinnamon and black pepper and was originally introduced to Scotland following the return from France of Mary, Queen of Scots. It was previously called Scotch Christmas Bun.

Tradition says that a bean should be hidden in the cake and the lucky finder becomes king – or queen – for the evening. Mary enjoyed these games. Once, in 1563, her companion Mary Fleming discovered the bean in the Black Bun so Queen Mary dressed her in royal robes and jewellery and she became queen for the evening. The English ambassador, however, did not see the funny side of the incident and wrote, 'The Queen of the Bean was that day in a gown of cloth of silver, her head, her neck, her

Black Bun. (© IMBJR under Creative Commons 2.0)

shoulders, the rest of her whole body, so beset with stones, that more in our whole jewel-house was not to be found.'

Then all the frivolity came to an end when Christmas was outlawed in Scotland after the Scottish Reformation in 1560. No more beans in Black Buns, no more royal charades.

> It has been suggested that Black Bun is simply a bigger version of a Garibaldi biscuit. Could it be that the biscuit originated because of the black bun? Garibaldi inventor John Carr was also Scottish.

HP Sauce

Aficionados of this fruity, rich sauce have **Frederick Gibson Garton** to thank because he was the one who invented and developed it. Frederick, a Nottingham grocer, registered the

Houses of Parliament. (© VisitBritain/BritainonView)

name in 1895. He called it 'HP' because he had heard that a restaurant in the Houses of Parliament had started to serve it, and for many years bottle labels have been emblazoned with a picture of the Houses of Parliament.

Garton sold the recipe and HP brand to Edwin Samson Moore for £150 plus the settlement of some unpaid bills. Moore, founder of the Midlands Vinegar Company, the forerunner of HP Foods, subsequently launched HP Sauce in 1903.

HP Sauce became known as 'Wilson's gravy' in the 1960s and 1970s after Harold Wilson, the Labour Prime Minister. The name arose after Wilson's wife, Mary, gave an interview to *The Sunday Times* in which she claimed, 'If Harold has a fault, it is that he will drown everything with HP Sauce.'

Tin Cans

Tins of beans, tins of soup, tins of vegetables – we tend to take them for granted. The cans, that is. But who knows what we would do if **Peter Durand** hadn't patented them in 1810? Okay, so Frenchman Nicholas Appert first preserved food by packing it into glass jars and cooking it for hours to sterilise it, but Peter Durand adopted the same method with the tin can. Great idea, but Peter hadn't really thought it through that carefully, because the tin cans were so thick a hammer and chisel were required to open them. The tin opener wasn't patented until 1855.

In 1813 John Hall and Bryan Dorkin opened the first commercial canning factory in England. Then, in 1846, Henry Evans invented a machine that could manufacture tin cans at a rate of sixty per hour, which was a significant increase over the previous rate of only six per hour.

Bottling was an early precursor to canning. Vegetables and fruit would be cooked in brandy, oil, brine or vinegar and sealed in bottles.

Sparkling Wine/Bubbly

Ah, surely it is quintessentially French? Well, no, because claims suggest that it was an Englishman who invented bubbly. West Country scientist **Christopher Merret**, born in 1614, used his knowledge of making cider to devise two techniques fundamental to making fizz. That was more than thirty years before French Benedictine monk and cellar master Dom Perignon.

For almost three hundred years, the French have made sure that only sparkling wine made in a particular part of France

Champagne.
(© Food4Media)

can be given the name 'champagne', but there are those who are certain that Merret recorded a recipe for a champagne-style drink twenty years before Perignon. If the truth be told, the French apparently preferred their white wine still; it was the English who got a taste for bubbly.

In December 1662, Dr Merret presented the Royal Society with a paper detailing the experiments of English cider makers. They had started to add sugar to wine to create a bubbly, refreshingly dry drink, similar to modern-day champagne. Dr Merret noted how 'our wine-coopers of recent times add vast quantities of sugar and molasses to wines to make them brisk and sparkling'.

The learned doctor detailed a second fermentation process – a chemical reaction that occurs when the bottled alcohol undergoes an increase in temperature and produces carbon dioxide – that now forms a key element in making champagne, known as the *methode champenoise*.

Of course, the French insist that Champagne was invented in 1697 by Perignon. He supposedly discovered it by accident; the wine he bottled from the abbey's vineyards in autumn just before the weather turned cool never fermented properly, only doing so when temperatures started rising again in the spring. Quite often bottles exploded due to this secondary fermentation thanks to the wine's dormant yeast producing sudden carbon dioxide bubbles. At first Dom Perignon viewed this as a problem, calling the drink 'the devil's wine' because one exploding bottle would often cause another to blow up, occasionally shattering entire cellars of wine. When he tasted the alcohol produced in bottles which didn't explode, though, he began to experiment with different varieties of grapes and came to see that it was a delicious fizzy drink.

However, there are wine experts who believe the French copied Merret's formula after visiting England. Perhaps it will never be proven, though Tom Stevenson, author of Christie's *World Encyclopaedia of Champagne & Sparkling Wine*, who has researched Dr Merret's work, says, 'We definitely beat the French and Dom Perignon by at least twenty-two years.' Whether we did or didn't, there's no denying that French champagne is hard to beat …

Anyway, on the English side of the Channel winemakers had also started trying to produce bottles which could withstand the secondary fermentation pressure, with some help from Dr Merret. One of his many disciplines included glassmaking. He was helped by advances in British furnaces, which were using coal as the fuel of choice instead of charcoal, after the British Navy requisitioned much of the timber to build a more powerful fleet. This allowed much higher furnace temperatures, which in turn facilitated the creation of stronger glass. Merret did not invent the technique, but he logged the process and allowed other winemakers to use it. Did you know that a bottle of champagne at room temperature contains about 49 million bubbles?

> The queen served sparkling wine from Chapel Down, Kent, at the wedding of the Duke and Duchess of Cambridge. What's more, the royal family has planted 16,700 chardonnay, pinot noir and pinot meunier vines in Windsor Great Park with the intention of creating their own fizz.
>
> Marilyn Monroe is said to have once taken a bath in champagne from 350 bottles.

Pimm's

Of course, not everyone likes bubbly. Some prefer other alcoholic beverages, like Pimm's for instance.

Kent man **James Tee Pimm** was educated in Edinburgh, where he studied theology. He then relocated to London and set himself up as a purveyor of shellfish. In 1823 he opened an oyster bar in Poultry Street. By 1833, this high flyer had opened another four restaurants. James created Pimm's No. 1 to compliment the flavours of shellfish and to provide a digestive aid. It was based on gin, quinine and a secret herb mixture. At that time, it was sold by the pint in pewter tankards.

> Over the years the Pimm's range was extended to include spirits other than gin as bases. So, whisky is the base for Pimm's No. 2, brandy is the spirit of choice for No. 3, rum for No. 4, rye whiskey for No. 5 and vodka for No. 6.

Corkscrew

A corkscrew, that essential device needed to open most wine bottles – one of its primary functions – has a screw at one end. This works its way down into the cork and grips it, providing a means by which the cork can be removed.

Sealing bottles with corks is not a new idea. This goes back to the days of the ancient Greek and Roman civilisations. At that time removing corks was not difficult because they extended above the rim of the bottleneck far enough to be grasped firmly. As time went on and bottles started to be mass produced, cylindrical corks were compressed before being forced into the bottle necks and were harder to remove than the earlier, tapered versions.

Fortunately, the corkscrew was invented around the same time. The first known patent for this essential tool was granted in 1795 to a vicar, **Samuel Henshall**, who came from Middlesex. It was a T-shaped device with a steel worm protruding perpendicularly from the centre of a handle made of bone or wood. Not long after that, people started to experiment with different designs for the same object.

In 1802 a more complex corkscrew was patented by **Edward Thomason**, who took over his father's business in Birmingham and applied for several patents. Thomason was confident that his invention was a winner, and every patent badge applied to the corkscrews produced in his factory carried the words 'Thomason Patent' beside the words '*ne plus ultra*', meaning 'no more beyond'. In fact there *were* more beyond – in 1888 **James Heeley**, also of Birmingham, invented a double-winged corkscrew, a type with levers on either side that is still used today.

During the latter half of the eighteenth century, corkscrews became ever more elaborate and were fashioned from silver, gold, exotic wood and jewels. Multi-purpose tools would combine corkscrews with devices including pipe nutmeg graters, seals and folding pocketknives.

Anaglypta

The invention of Anaglypta by **Thomas John Palmer** is closely linked to the invention of linoleum by Frederick Walton (see next entry).

The manufacture of wallpaper was originally a labour-intensive process and therefore could be afforded only by the wealthy. The automated manufacturing process introduced in 1841 made it much more affordable and available to the mass market.

In 1877 Walton invented a wall covering called Lincrusta, similar in many ways to linoleum. It was popular because it was the first washable wall covering, decorative and highly durable. Its composition meant, however, that it was both heavy and inflexible and in time could harden. Nevertheless, it is still available today from specialist shops.

One of Walton's employees, Thomas Palmer, created a similar but lighter product made from wood pulp and cotton in 1883. Unlike Lincrusta, it could also easily be painted. Walton realised his Lincrusta product could not compete with the lighter and more flexible invention and decided not to patent it. Palmer therefore patented his product, left Palmer's employment and in 1886 moved from west London to Lancaster where, in partnership with a local firm, Storey Brothers, he started producing Anaglypta in 1887.

However, in 1894 production was moved to Darwen, where the continuous manufacture of wallpaper by machine had been developed at Potters & Ross, a cotton printing company. Such was the success of the process that by the time Anaglypta was first being produced, the Potters were renowned around the world as suppliers of wallpaper. With the resources of the Potters supporting the Anaglypta company, its success was assured.

In 1899 the Anaglypta company was acquired by The Wall Paper Manufacturers Ltd, and the following year the product was awarded two gold medals at the Paris Exhibition of 1900.

The name Anaglypta is derived from the Greek words *ana* (raised) and *glypta* (cameo).

Linoleum

Around 1860, a floor covering called Linoleum was invented by **Frederick Walton**. He realised linseed oil could be made into a waterproof material and that if he applied the varnish to a backing, he could sell it as a ready-made floor. Thus linoleum was born.

Frederick was born near Halifax in 1834, and when he was twenty-one he entered into partnership with his father and elder brother. Frederick had his own workshop and experimented and produced clothes brushes, hairbrushes and horse brushes based on the use of India rubber.

During this time he noticed that a rubbery, flexible skin of oxidised, solidified linseed oil had formed on a can of oil-based paint and it occurred to him that he could use such a skin to waterproof materials, similar to how rubber was used. He managed to produce a quantity of oxidised oil and discovered that it had some of the properties of rubber.

His brush business was not a financial success, so around 1861 he set off for London, settling in Chiswick.

A rubber-based floor covering, Kamptulicon, was available but was not cost-effective due to the rising cost of rubber. Also, unlike his own material, it could not be rolled on to a backing, whereas his own product could be rolled on to a backing in a single process. This process was later to lead to the mass production of linoleum.

In 1861 he took out a patent for the manufacture of varnish applicable to the waterproofing and coating of fabrics and other uses. He filed his patent in April 1863. To quote the patent document, 'this invention has for its object improvements in the making of fabrics for covering floors and other surfaces ... canvas or other suitable strong fabrics are coated over their upper surfaces with a composition consisting of oxidised oil, coal dust and gum or resin, preferring Kauri or New Zealand gum, such surfaces being afterwards primed, painted, embossed or otherwise ornamented'. Coal dust was replaced by ground cork but the basic idea for linoleum remained as stated in the patent.

Initial production took place at his home in Chiswick but the inflammable process resulted in serious damage, so around 1864 he moved production to an old calico mill in Staines, west London, which already had the rollers he needed for large-scale production.

His successful company, the Linoleum Manufacturing Company, produced a range of products based on his patent and such was the demand that they were exported as far afield as Europe and the USA. Scottish manufacturer Michael Nairn developed the inlaid patterning that became popular in linoleum.

Walton also created a washable wall covering called Lincrusta, but it suffered from weight and lack of flexibility. One of Walton's employees, Thomas Palmer, invented a similar product using wood pulp and cotton, and this product was lighter and more flexible. As a result, Palmer left Walton, patented his product, and Anaglypta was born (see previous entry).

The product was originally called Kampticon but it soon became apparent that this was confusingly similar to the inferior rubber-based product Kamptulicon. The product was renamed 'linoleum', based upon the Latin '*linum*', meaning flax, and '*oleum*', meaning oil.

Collapsible Baby Buggy

Former test pilot and aeronautical engineer **Owen Maclaren** invented this in 1965. He was the same man who helped design the Spitfire's folding undercarriage during the Second World War.

The idea for the buggy was born when the Northamptonshire man saw his daughter struggle up some steps with a heavy pram. He thought there must be an easier way to take a child around, and set to work to solve the problem.

Familiar with lightweight, durable materials from his work with the Spitfire fighter plane, he adapted his work to create a stroller. His idea was to make one that could provide his granddaughter with comfort and safety without being cumbersome and heavy. His first model of the buggy had a frame of tubular aluminium, so it was light and folded up neatly. In 1967 the item became hugely popular and Owen Maclaren set up a factory in Long Buckby to manufacture what he called the Maclaren B-01.

Today, a modern version of his lightweight, foldable buggy is sold in more than fifty countries and has been used by millions of people worldwide, including royals and celebrities.

Early baby carriages were enormous affairs. They were often wooden with brass fittings. Some were named after royalty, and Princess and Duchess were popular models.

The Waterbed

Born in 1788, Scotsman **Neil Arnott** from Arbroath in Angus was a natural philosopher and physician to the French and Spanish embassies. In 1837 he became physician extraordinary to the queen, and he was one of the founders of the University of London.

Although his inventions included an efficient smokeless stove (which he called Arnott's Stove) and a ventilating chimney valve (the Arnott ventilator), his biggest claim to fame came from his Arnott hydrostatic waterbed, which he invented in 1832. The bed consisted of light bedding placed on top of a rubberised canvas, which rested on a bath of water. The waterbed allowed mattress pressure to be evenly distributed across the body, thereby helping reduce bedsores in patients. However, there was one fault inasmuch as his invention could not regulate the water temperature. Arnott did not patent his invention, leaving it to other physicians to refine and rework his basic design.

Portsmouth's Dr William Hooper was also well aware of the medicinal benefits of waterbeds, and patented his waterbed invention in 1883. Unfortunately his idea was also a non-starter, since he too could not quite figure out how the temperature of the water should be regulated.

Meanwhile, Harrods thought the concept was a great idea. This would be a moneymaker and no mistake. They set out to sell waterbeds as a luxury mail-order item in 1895. Unfortunately, the material used to make the beds was practically the same as that used for large hot-water bottles. Back to the drawing board. This was another defect that let down the item's commercial potential.

Then, along came the 1960s. Vinyl was invented and suddenly waterbeds were the tops. No longer were they regarded as a peculiar luxury item. Now they were a practical commercial product, complete with heaters and

Harrods. (© BritainonView)

liners. Nowadays, the waterbed combines the medicinal benefits of pressure reduction and comfort with the highest standards of luxury.

> Although waterbeds had their heyday in the late 1960s, the earliest known form of a waterbed dates back over 3,600 years to Persia. Goatskins filled with water were placed in the sun to absorb heat. Unlike their historical predecessors, modern waterbeds offer sleepers the ability to control the temperature of the water without having to first leave them in the sun.

Refrigeration

Can you imagine your kitchen without a fridge? It is true to say that this invention has had a massive impact on today's world, radically improving our way of life.

At room temperature bacteria multiply rapidly, making food inedible and sometimes even dangerous. In cold temperatures, bacterial activity slows dramatically before, at freezing point or below, it stops. Keeping food cold therefore allows it to maintain its freshness for days when otherwise it might last only hours. Freezing can preserve food for months or years. Those brainy folks in ancient Greece and Rome knew this and stored snow in insulated pits for exactly this reason. Centuries later people still kept food outside, for example in storage tins buried in the ground. Of course it was not really efficient, but it was better than nothing.

Some foods were preserved by methods such as drying, pickling or smoking, but that was only suitable for a small range of produce.

Enter **Dr William Cullen** of Glasgow University, who invented the first method of refrigeration – cooling air by the evaporation of liquids in a vacuum – in 1748. By means of a pump, he sucked air out of a bell jar and created a partial vacuum over a container of a colourless liquid. This was known as 'diethyl ether'. It came to a boil as depressurisation continued and absorbed heat from the nearby air, which in turn became so cool that a small amount of ice was formed. Cullen did not apply his discovery to any practical purposes but fortunately others eventually realised that this was a really 'cool' idea. (Sorry.)

Refrigeration meant that, for the very first time, families with no means of growing their own food could enjoy a more varied diet, and fresh fruit and vegetables became staples. People could sample exotic and previously unknown produce, which was imported from abroad using refrigeration on ships. Shoppers were able to stock up, making savings by buying in bulk and storing their goods for long periods. Milk could be kept for days. Primitive methods of keeping food cool, some little altered from Roman times but still used in 1930s Britain, were finally consigned to history.

Frigidaire. (©
Gilly Pickup)

In the early 1950s most refrigerators were white. In the 1960s, pastel colours such as turquoise and pink became trendy and brushed chrome-plating, similar to stainless finish, was available on some models from different brands. In the 1980s black became fashionable, and in the late 1990s stainless steel came into vogue.

Meat and fish immersed in barrels or tubs of brine were used to feed Nelson's navy. This had a terrible taste and was very unhealthy. Only the best meat was salted, and that's where the expression 'not worth its salt' comes from.

Oil-Powered Cleaner/The Puffing Billy

This was invented in 1901 by Gloucester-born engineer Hubert Cecil Booth. One day, when he was watching a railway carriage being cleaned by a machine that blew the dust away, an idea struck. Why blow the dust away? Why not create an oil-powered machine that sucked the dust up instead? To test his theory, he placed a handkerchief on a chair and sucked through it, finding that dust collected on either side. Booth wasted little time and first demonstrated his vacuuming device in a restaurant that same year, showing how it could suck up dust.

He set up a cleaning service using a large horse-drawn, petrol-driven unit, which was parked outside the building to be cleaned, with hoses stretching through the windows. Booth patented the motorised vacuum cleaner on 30 August 1901. This was considered quite extraordinary; no one had seen anything like it before. Before long, he was asked to perform unusual jobs; one was to clean the girders of Crystal Palace, which were thick with layers of dust accumulated over the years. He didn't only send one machine to do the job as he was asked; he sent fifteen of them. Over the four weeks it took to complete the task, over twenty-six tonnes of dust had been vacuumed up.

His vacuum cleaner was popular with the rich, and even started to appear in royal households. By 1903 society ladies were having vacuuming parties, an early equivalent of tupperware or underwear parties. They invited their friends over to watch while workers used Booth's machine in their homes. Hoover came along in later years and made it into a convenient upright. Ladies have thanked Booth for his idea ever since, and even more so nowadays ...

... With the Bagless Vacuum Cleaner

Though the original idea is over a hundred years old, inventors keep coming up with advanced ideas. Take **James Dyson**, for example, the definitive British technical innovator. The Cyclon, his dual-cycle bagless vacuum cleaner, took him over fifteen years and more than 5,000 prototypes to perfect, but became the fastest-selling vacuum cleaner in British history. It took guts and perseverance and many would have given up, but he was sure he was onto a winner even when in 1982 Hotpoint told him, 'This project is dead from the neck up.'

In the late 1970s, Dyson had the idea of using cyclonic separation, which uses a spinning cone to separate the dirt from the air, to create a vacuum cleaner that would not lose suction as it picked up dirt. After another five years and 5,127 prototypes, Dyson launched the G-Force cleaner in 1983. It was bright pink in colour and sold at £2,000. It won the 1991 International Design Fair prize in Japan. Dyson obtained his first US patent in 1986.

Unfortunately, no manufacturer or distributor would launch his product in the UK as it would disturb the valuable cleaner-bag market. Not one to be easily deterred, Dyson set up his own manufacturing company. In June 1993 he opened

A 1980s magazine advertisement for the
Rotork Cyclon bagless vacuum cleaner

Cyclon. (© Frenchay Village
Museum)

his research centre and factory in Malmesbury, Wiltshire. The product now outsells those of some of the companies that rejected his idea and has become one of the UK's most popular brands.

In early 2005 it was reported that Dyson cleaners had become the market leaders in the US by value, while the Dyson Dual Cyclone became the fastest-selling vacuum cleaner ever to be made in the UK.

Dyson's breakthrough in the UK market, more than ten years after the initial idea, was through a TV advertising campaign which emphasised that unlike rival vacuums, with his there was no need to buy replacement bags. At that time the UK market for disposable cleaner bags was £100 million. The slogan 'Say goodbye to the bag' proved more attractive to the buying public than a previous emphasis on the suction efficiency.

The Dyson 360 Eye is the firm's first robotic cleaner, the product of £28 million in research over sixteen years. This baby has a panoramic camera that takes thirty images every second so it knows where it has been and where to go next, using infrared sensors to move on tank-like tracks that negotiate obstacles. How clever is that?

James Dyson explained, 'Most robotic vacuum cleaners don't see their environment, have little suction and don't clean properly.' This one, however, with its 360-degree camera and Dyson's suction system, is a high-performing and 'genuine labour-saving device'. Additionally – and surely this is the best part – you can control it with your mobile phone while you are out shopping so you return to a nice, clean house.

In 1997, Dyson was awarded the Prince Phillip Designers Prize. In 2005 he was elected as a Fellow of the Royal Academy of Engineering, and then he was appointed a Knight Bachelor in the New Year Honours, December 2006.

In the days when vacuum cleaner salesmen went from house to house to try and sell their wares, they would sometimes try to bully women into buying one of their machines by listing all the horrible things that household dust contained. So here we go: one teaspoonful of dust can contain 5 million skin cells, 10,000 dust mites, hundreds of pet hairs (if there was a pet in the house), insect legs and eyes and innumerable other creepy-crawly body parts, millions of bacteria, and grains of plant pollen.

Carbonated Water

Around 1772 clergyman and scientist **Dr Joseph Priestley** from Leeds – best remembered for his isolation and description of several gases, particularly oxygen – accidentally invented carbonated water when he suspended a bowl of distilled water above a beer vat.

Priestley lived next to a brewery and was apparently intrigued by the 'air' that floated over fermenting grain. His experiments showed that this 'heavier-than-air' gas was able to extinguish burning wood chips. This gas would later be identified as carbon dioxide. Priestley devised a new way to produce this 'heavy gas', as he called it, in his home laboratory. He poured acid onto chalk (calcium carbonate), and heavy gas resulted. On dissolving this gas in water he found that it had a pleasant and tangy taste. Soda water had just been invented! His invention of carbonated water is the major and defining component of most soft drinks. These carbonated drinks were often flavoured with lemon, making them the forerunner of lemonade.

He published a paper describing how to make carbonated water with the impressive title *Impregnating Water with Fixed Air*, and just a few years later Jacob Schweppe set up a London factory and began manufacturing fizzy drinks using Priestley's method. Interestingly, Schweppe, never one to miss a trick and the first person to come up with a machine to add fizz to water, sold it as a cure for indigestion and gout.

Priestley didn't stop at carbonated water, though; he also discovered oxygen, hydrochloric acid, nitrous oxide (laughing gas), carbon monoxide and sulphur dioxide, in addition identifying plant respiration and photosynthesis.

For his soda water invention Priestley was elected to the French Academy of Sciences in 1772, and the following year he received the prestigious Copley Medal from the Royal Society.

Jigsaw Puzzle

If you are a fan of jigsaw puzzles, you have **John Spilsbury**, a Worcester-born cartographer and engraver, to thank. Around the 1750s Spilsbury was apprenticed to a London engraver and map seller called Thomas Jeffreys who at that time was Geographer to the King. After a few years spent learning the trade, Spilsbury went into business for himself.

A 1763 street directory lists him as 'Spilsbury, John. Engraver and Map Dissector in Wood, in order to facilitate the Teaching of Geography. Russel-court, Drury-lane.'

In the late 1760s, he began gluing world maps on to hardwood veneers before cutting them along the borders of the countries and regions with a jigsaw to create separate pieces. The end product was an educational tool to teach geography. Originally the puzzles were called 'dissected maps' or 'dissections', and the name 'jigsaw' didn't come into being until much later.

The idea soon caught on, and until about 1820 jigsaw puzzles remained, first and foremost, tools for learning. The pieces in these early puzzles were not interlocking. It wasn't until the invention of power tools more than a century later that jigsaw puzzles with fully interlocking pieces come into being.

Before long, people began making pictorial jigsaw puzzles to entertain rather than teach. Subjects became wider-reaching, and soon puzzles did not represent just maps but biblical scenes, popular stories of the day and poetry.

Towards the end of the century, plywood came to be used as the puzzle base. With illustrations glued or painted on the front of the wood, pencil tracings of where to cut were outlined on the back. Some of these tracings can be found on some old puzzles today.

It was not until the twentieth century that cardboard puzzles came to be die-cut, a process where thin strips of metal with sharpened edges are twisted into complex patterns and fastened to a plate which is pressed down on the cardboard to make the cut. Wooden puzzles still dominated, though, as manufacturers were confident that customers would not be interested in cardboard puzzles. Of course, jigsaw manufacturers and retailers were also driven by profit – and wooden puzzles were more expensive than cardboard ones.

The golden age of jigsaw puzzles came in the 1920s and 1930s, when British companies such as Chad Valley and Victory started to produce a range of puzzles.

The largest commercially available jigsaw in the world listed by the *Guinness Book of Records* measures 13 feet by 7 feet.

And to complete this section, here are some more Great British Inventions in the domestic category ...

Kendal Mint Cake

If you are a hiker or climber, you may have packed a bar or two of this glucose-based confection flavoured with peppermint in your rucksack. Originating from Kendal in Cumbria, the bar is a source of energy. It originated from a batch of peppermint creams that went wrong after the mixture was left overnight and the solidified 'mint cake' was discovered in the morning. Edmund Hillary and his team carried Romney's Kendal Mint Cake with them on the first successful ascent of Mount Everest in 1953. The packaging currently includes the following: 'We sat on the snow and looked at the country far below us ... we nibbled Kendal Mint Cake.' A member of the successful Everest expedition wrote, 'It was easily the most popular item on our high-altitude ration, our only criticism was that we did not have enough of it.'

Piccalilli

The *Oxford English Dictionary* traces the word 'Piccalilli' to the middle of the eighteenth century when, in 1758, English cookery writer **Hannah Glasse**, best known for her cookbook *The Art of Cookery*, first published in 1747, described how to make 'Paco-Lilla' or 'India Pickle'. The spelling we are now familiar with, 'piccalilli', was used in an advertisement in a 1799 edition of *The Times*. British piccalilli contains cauliflower and vegetable marrow and seasoning of mustard and turmeric. Piccalilli is popular as a relish eaten with cold meats or with a ploughman's lunch.

Dewar Flask

Invented in 1892 by **Sir James Dewar**, but hardly anyone calls it a Dewar Flask so it may be more recognisable when given its popular name, the Thermos flask. Dewar, who was born in Kincardine-on-Forth, Scotland, came up with this bright idea when he was a professor of chemistry at Cambridge. Not that

Thermos brand vacuum flask.
(© Denae Bedard under Creative
Commons 2.0)

he invented it to keep his tea hot on picnics – he probably never went on one – but to help his research on cooling gases, like air and oxygen, to such low temperatures that they would liquefy. The resulting oxygen machine saved lives.

The Enamel Bathtub

The Scotsman behind Buick cars (see the section on movement) is the same who invented the enamel bathtub. **David Buick**, who was born in Arbroath, Angus, went to live in Detroit, Michigan, and worked for a plumbing firm where he invented the process of permanently bonding porcelain enamel to cast-iron baths. Although cast-iron baths are rarely seen nowadays, the method, a key step in the history of manufacturing, is still in use for enamelling them.

2

Movement

Lawnmower

In 1827 mechanic **Edwin Beard Budding** of Stroud, Gloucestershire, was considered crackers for inventing a contraption as weird as this. In fact, rumour has it he had to test the machine at night so no one would see him. His inspiration came from cross-cutting machines that finished woollen cloth in a local mill – he repaired machinery for the textile mills of the Stroud valleys – and he figured that a similar tool would be effective in trimming grass.

Before the invention of the lawnmower, lawns were usually scythed, or kept neat and tidy with shears or a sickle, a laborious and highly skilled task if a short, even finish was to be achieved.

Budding's lawnmower, a reel-type affair with a series of blades arranged around a cylinder and the world's first of that type, was classed as a 'Two Man Mower – One Pushing One Pulling'. Adverts of the day claimed, 'Gentlemen will find my machine an amusing & a healthy exercise plus do the work of 6 men.'

Budding's mower was based on a simple but effective design. A series of linear, gently curving blades fixed to a horizontal cylinder passed over a fixed bottom blade. As the mower pushed forward, the rotation of the rear roller drove a series of gears that spun the cylinder at much-increased speed, ensuring a high rate of cuts as the grass was flicked up into its path. The combination of spinning cylinder and rear roller had the added benefit of leaving a rather nice stripe on the lawn. The first machine produced was nineteen inches wide with a frame made of wrought iron. The mower was designed primarily to cut lawns on sports grounds and extensive gardens, and on 31 August 1830 he patented 'a new combination and application of machinery for the purpose of cropping or shearing the vegetable surface of lawns, grass-plats and pleasure grounds'.

These machines were the worldwide catalyst for the preparation of sporting ovals, playing fields, pitches and grass courts and this also led to the codification of modern rules for many sports, including most football codes, lawn bowls and lawn tennis.

In one of Jane Webb Loudon's gardening books, *Instructions for Gardening for Ladies* (1840), she commended Budding's invention, saying, 'It is particularly adapted for amateurs, offering an excellent exercise to the arms and every part of the body.'

However … an Essential Addendum to the above Story…
Scotsman W. F. Carnegie from Arbroath was another who purchased a Budding lawnmower. There was a problem, though, because Carnegie's lawns covered two and a half acres and he felt that the Budding machine was not capable of doing the job efficiently and quickly enough.

So Carnegie engaged a local engineer, **Alexander Shanks**, whose firm was founded in Arbroath in 1840, originally producing iron castings, steam engines and excavating machinery. Carnegie asked him to make a twenty-seven-inch-wide machine which could be pulled by two men or a pony. Subsequently the pony got most votes as it left no traces on the grass when cutting was carried out in dry weather – ponies often wore leather slippers to prevent hoofmarks. When the pony pulled at the front the operator normally walked behind, although a few designs provided a seat above the rear roller. Although it ought to have been possible for the mower and pony to be controlled by one person, most contemporary photographs show a second person, possibly the gardener's apprentice, walking alongside the pony to guide it.

Shanks started lawnmower production in 1842 and patented his design in Scotland for a lawnmower that could cut grass and roll the turf in one operation. Up to 1852

Advertisement for 'Shanks Patent Lawn Mowers' for 1866.

Scotland had its own patent system, so Budding's patent only covered England and Wales.

Although Alexander Shanks died at a young age, his son James exhibited the mowers at the Great Exhibition in 1851. This led to a successful business that continued into the twentieth century and they supplied mowers to cut the tennis courts of Wimbledon, Lord's Cricket Ground and the Old Course at St Andrews.

In 1846 a machine was sold to Clumber Park in Nottinghamshire and an article written at the time read, 'The machine has been in constant use in the gardens at Clumber for upwards of three months. It is constructed on the same principle as Budding's patent mowing machine but altogether stronger and, of course less liable to go out of repair, the cutters are forty-two inches in length, it is drawn by one horse requiring a boy to lead the horse and a man to direct the machine. The saving in labour has amounted to seventy per cent.'

Pony and horse mowers continued in popularity for many years after the introduction of steam- and petrol-powered mowers. Some were still in use up to the 1940s, although after the Second World War the motor mower had replaced most of them, especially in Europe and North America.

It may be fanciful to think that this is where the term 'Shanks's Pony', meaning 'to walk', originates, but it certainly makes sense if one takes the phrase to mean 'having a pony but still having to walk'.

Catseyes

It was a dark and rainy night in 1933 when road contractor **Percy Shaw** from Halifax was driving home. He saw his car headlights reflected in a cat's eyes. So one story goes. However, in a television interview with revered traveller and broadcaster Alan Whicker, Percy told a different story of being inspired on a foggy night to think of a way of moving the reflective studs on a road sign to the road surface. Whichever is the true story of Percy's eureka moment – and presumably it is the second – he knew if he could create a gadget that replicated the effect, he could make driving at night safer for everyone.

Not that it was the first time that Percy had showed inventiveness. Born into a poor family, with ten older brothers

Lightdome Road Stud. (© Eliot2000 under Creative Commons 2.0)

and sisters and two younger ones, all the children had to help with the chores. Their father was a labourer who struggled to support his family on a £1 weekly wage from a local textile mill. They grew fruit and vegetables in their garden and Percy would sell the surplus in the village. He also grew herbs which he sold at a penny a bunch. He left school at thirteen to work as a labourer. He knew that if he was to have any kind of a future he would have to better himself, so he went to classes to learn shorthand and bookkeeping and before too long managed to find himself a job in a factory office.

His quick thinking earned him some much-needed cash when one day he prevented a factory fire from spreading by running outside to cut off the general gas supply. His action meant he had cut off the gas for the whole village, but he was given a reward of 7/6d, more than a week's wages at the time.

He decided it may be advantageous to learn welding and machine-tool making too. For a time he worked in a carpet factory where he invented a way to provide rubber backings, the forerunner of foam-backed carpets. However, his idea came to nothing. He also tried to design a petrol pump, but that too fizzled out before it began.

When cars began to increase in popularity, Percy repaired them when they broke down and invented a way to re-metal big ends. This alleviated the need to send the faulty item back to the manufacturer and take the car off the road for some time.

As time passed, Percy bought a bicycle and cycled to London with two friends, a journey which took them three days. It was his first holiday. By 1916 Percy had saved enough money to buy a second-hand motorbike and then a Ford Model T. He started a business to lay asphalt and tar macadam on private roads and garden paths. He found that the small roller used to compact them was slow, so set about inventing a miniature motorised version. Soon he was employing several men to do the work while he travelled round Yorkshire securing orders.

So all because of driving home on that dark, rainy night, reflecting road studs were born. The original catseye works by housing a pair of glass 'eyes' in a white rubber casing which is laid into the road's surface. The 'eyes' reflect light from a vehicle's headlamps back so that the driver can see the road ahead. The original catseye is strong, with the casing dipping into the road when a car runs over it. A built-in rubber wiper cleans the glass eyes, so helping them to keep their visibility.

Not that Percy took a break from his work. He kept on improving his design. He even dug up a road without permission, installed a road stud, tried it out with his own car headlights, reinstated the road, and took the stud home to work on it.

After attaining the patent in 1934, he set up his company 'Reflecting Road-Studs Ltd' and successfully manufactured his product until his death in 1976. Sales of catseyes were initially slow, but Ministry of Transport approval, followed by wartime blackout regulations designed to thwart German bombers, sent demand rocketing sky high. However, there was a problem when in 1941 the Japanese Army entered Malaya and supplies of rubber plummeted. Percy's company was allowed only a small quota. Scrap rubber was found to be ineffective but old crepe soles, treated with alcohol and petroleum, did the trick, although production at that time fell to 12,000 a week.

After the war, Percy became famous. Parliament described the road stud as 'the most brilliant invention ever produced in the interests of road safety'. In July 1949 the queen, who was then Princess Elizabeth, visited Halifax, saw a display of road studs and asked to meet the inventor. It wasn't long before more than a million road studs a year were being made and Percy became a very wealthy man.

In later life, old Percy became something of an eccentric by all accounts. Visiting the patent attorney's office one day, he was asked to take away samples of earlier road studs and a new recruit was told to carry them down to his car. Struggling to open the boot of a Morris Minor parked outside the office, Percy arrived and said to him, 'Nay lad, that's not car, that's car over t'other side.' Across the road was a Rolls-Royce complete with uniformed chauffeur. Percy then went to the market to buy tripe for his lunch, which he ate with his fingers in the back of his Rolls. He didn't much care for home comforts but he had his cellar full of crates of Worthington's White Shield, his favourite beer.

He was awarded the OBE for services to exports in 1965. Road Reflecting Studs Ltd prospered and is still located in Boothtown, Halifax. It ships over a million road studs worldwide each year.

It is worth remembering that even the simplest idea can have a profound and lasting significance.

Percy's catseye was voted the greatest invention of the twentieth century.

Belisha Beacon

They make it easier and safer to cross busy roads when there are no traffic lights around. This marker of pedestrian crossings was named after the mellifluously named **Leslie Hore-Belisha**, Minister of Transport and the man responsible for instituting military conscription in 1939, a few months before the outbreak of the Second World War.

The beacon, an amber-coloured globe on top of a tall black-and-white pole, became a familiar sight in towns and cities across Britain in 1934. At first, the crossings were marked out by large metal studs in the road, but later they were painted in black and white stripes and became known as 'zebra crossings'. Belisha beacons gave extra visibility to zebra crossings for motorists, particularly at night.

These days, many zebra crossings and Belisha beacons have been replaced by pelican crossings or puffin crossings because when it comes to Belisha beacons pedestrians just have to hope that the traffic will stop for them. With pedestrian-controlled traffic signals, a waiting pedestrian can stop traffic by pressing the button and waiting for the pedestrian signal of a red man to change to green. The green man is sometimes accompanied by a green bicycle to indicate that the crossing is designated for pedestrians and cyclists which, continuing with the bird theme, is called a Toucan crossing, as in, 'two can' cross. Another variation is the pegasus crossing, where the pedestrian is accompanied by a green horse to indicate that the crossing is designated for pedestrians and horses. One of these crossings can be seen at London's Hyde Park Corner.

The first Belisha beacon became operational on 4 July 1935 in Wigan, Lancashire.

Bicycle

In the beginning, the bicycle evolved from a little wooden horse with a fixed front wheel, a French invention from the 1790s. It had no pedals and the fixed front wheel meant it could not be steered, so it was in effect a walking machine. Perhaps not surprisingly, it did not prove popular. In 1817 a German baron, Karl von Drais de Sauerbrun, improved on it by making two same-size inline wheels, the front one steerable to replace the fixed one mounted in a frame which the rider straddled. The device was propelled by pushing the feet against the ground, thus rolling yourself and the device forward in a kind of gliding walk. He named his invention the 'Draisienne', though generally people referred to as a 'dandy horse' or 'hobby horse'. It was made entirely of wood. This enjoyed short-lived popularity though was not at all practical for transportation anywhere other than a well-maintained pathway in a park or garden.

In 1839, there were further improvements when **Kirkpatrick Macmillan**, a Scottish blacksmith, introduced the first pedals. Macmillan had set out to make himself a hobby horse but had ideas for improvement. He knew that the machine would be much more useful if the rider's feet did not touch the ground so he worked on adding cranks and metal rods to turn the back wheel.

Macmillan had no particular desire to make commercial success from his invention. He didn't patent his new design for the 'velocipede' ('fast foot'), or 'boneshaker' as it was more commonly known, the latter of which names needs no explanation.

In 1846, **Gavin Dalzell** of Lesmahagow in Scotland produced a similar bicycle to Macmillan's. There were so many bikes made and sold around that time that for years Dalzell was considered the original inventor. His was a two-wheeled riding machine made entirely of wood with pedals applied directly to the front wheel. Later, metal tyres were to replace the wooden ones and these, plus the cobblestone roads, made for an extremely uncomfortable ride. The design was later improved by Henry Lawson and John Kemp Starley with the addition of a chain drive and gears. It was not until the 1860s that the bicycle really became popular, especially in France.

The next step came in 1870 when the high-wheel bicycle, the first all-metal machine, made its appearance thanks to

James **Starley**, who was born in Albourne, Sussex. It was called the penny-farthing and it was a strange-looking cycle. It was named after two coins of the day – the large penny, which the big wheel resembles, and the small farthing. It was thought that the bigger the front wheel, the faster the bicycle would travel.

Before this, metallurgy was not advanced enough to provide metal strong enough to make small, light parts. The pedals were still attached directly to the front wheel with no freewheeling mechanism. Solid rubber tyres and long spokes on the large front wheel provided a smoother ride than its predecessor. The front wheels gradually became larger as makers realised that the larger the wheel, the farther you could travel with one rotation of the pedals. You would purchase a wheel as large as your leg length would allow.

Since the penny-farthing had a hollow steel frame, solid rubber tyres and ball bearings it was lighter and therefore meant a smoother ride for the cyclist. It was only popular among young men of means, since it cost an average worker six months' pay. However, the penny-farthing wasn't without its problems. Because the rider sat so high above the centre of gravity, if a stone or rut in the road stopped the front wheel or a person suddenly appeared in the road, the entire machine rotated forward on its front axle and the rider, with his legs trapped under the handlebars, fell unceremoniously on his head. This is when the term 'taking a header' came into being.

While the men were risking their necks, ladies, restricted by long skirts and corsets, could take a spin around the park on the High Wheel Tricycle, a machine which also afforded more dignity to gentlemen such as doctors and clergymen.

Penny-farthing bicycle. (© Gilly Pickup)

Ironbridge World Heritage Festival. Costumed actors as Victorian policemen on penny-farthing bicycles. (© Visit England/Garreth Anderson)

Kirkpatrick was considered a character by the locals. They called him 'Daft Pete' and he was often seen going into Dumfries on his 'strange' invention. He was reported for speeding and once, so the story goes, knocked a little girl over, though she was unhurt.

So What about the Safety Bicycle?

Well, this was invented by Londoner **John Kemp Starley** in 1885. John was James's nephew and his invention was the forerunner of the bicycle as we know it today. It got its name because, without putting too fine a point on it, other bikes at the time – including the penny-farthing – were dangerous.

The further improvement of metallurgy meant that this particular machine, which became the model for the modern bicycle, had a chain, sprocket, driving rear wheel and equal-sized wheels with hard rubber tyres.

John knew the rider had to sit at the proper distance from the ground, that he had to place the seat in the right position in relation to the pedals and that the handles had to be positioned in relation to the seat that the rider could exert the greatest force upon the pedals with the least amount of effort. What made this bicycle the real deal was the chain drive, which meant the cyclist could ride fast even though both wheels were the same size. For most people, then, it was arguably the most liberating invention of all time.

So Starley designed the lightest, strongest, most reasonably priced, most rigid, most compact and ergonomically most efficient shape the bicycle frame could be.

The years that followed saw further improvements, with the development of pneumatic tyres and two- and three-speed gears. The bicycle had earned its place as a practical investment for the working man as transportation, allowing much more flexibility. Ladies, who up until then had been consigned to riding heavy adult-size tricycles that were only practical for taking a turn around the park, now could ride a much more versatile machine and still keep their modesty by wearing long skirts. The bicycle craze put an end to the impractical bustle and corset, instituted 'common-sense dressing' for women and increased their mobility considerably. It was said that the bicycle had done more for the emancipation of women than anything else in the world.

Bicycle. (© Gilly Pickup)

In 1896, J. K. Starley & Co. was floated as the Rover Cycle Company. The capital financed the construction of the largest cycle works in Coventry, then the global centre of bicycle manufacturing.

Pneumatic Tyres

Hurrah for vet **John Boyd Dunlop**, who came up with this idea in 1887. Actually, though, Dunlop from Ayrshire was not quite first with the idea, because in 1845 Scottish railway engineer Robert William Thomson invented the world's first pneumatic tyres. However, nothing came of his idea because there was no real market for them.

Forty-two years later, Dunlop came up with pneumatic or inflatable tyres after his eight-year-old son complained about getting headaches from riding his tricycle on bumpy surfaces. The boy didn't realise what his moaning would lead to, and after experimenting with his son's tricycle Dunlop came up with a design based on an inflated rubber tube in 1887. He patented the idea on 7 December 1888. This time round, the invention coincided with the new bicycle craze. That's what you call lucky timing.

However, it wasn't all good news, because two years after he was granted the patent Dunlop was informed that it was invalid as Thomson had actually patented the idea in France in 1846 and in the US in 1847. He had to fight a legal battle with Thomson, which he won.

John Dunlop did not benefit much financially from his invention; he sold the patent and company name early on. However, despite Thomson's work, today it is Dunlop who is accredited with inventing the first commercially viable rubber tyre. Within ten years it had almost entirely replaced solid tyres.

The company Dunlop founded went on to become the Dunlop Rubber Company.

Tarmac

Ask most people who invented tarmac to smooth the path of travel and they would be likely to say 'John McAdam', particularly if they are Scottish, like me. However, while Ayrshire-born McAdam invented the idea of crushed-stone road surfaces, he unfortunately failed to come up with an idea to make the stones stick. Fine though this was in the days of horse-drawn vehicles, the surface was unsuitable when cars started to appear. For one thing the jagged material meant tyres would puncture, and when it rained many roads became impassable due to mud.

Enter Nottinghamshire's **Edgar Hooley**. One day when he was out and about, he noticed that a stretch of road in Derbyshire had no ruts in it. When he queried this, he was told that a barrel of tar had accidentally fallen onto the road and waste slag from a nearby ironworks had been poured on top to cover up the mess. This accidental resurfacing had solidified the road and there was no rutting and no dust.

Hooley didn't hang around. By 1902, he patented the process of heating tar, adding slag to the mix and then breaking stones within the mixture to form a smooth road surface. Nottingham's Radcliffe Road became the first tarmac road in the world.

In 1903, Edgar Purnell Hooley formed Tar Macadam Syndicate Ltd and registered Tarmac as a trademark.

General Motors Corporation of America

David Dunbar Buick, who was born in Arbroath in Angus, founded the company that became known as General Motors Corporation of America, a mighty car-making empire.

His family moved to Detroit when David was a small boy, and when he was fifteen he went to work with Alexander Manufacturing Company, which made plumbing fixtures. Unfortunately the business failed in 1882, so Buick and William Sherwood bought it and renamed it Buick and Sherwood. Buick was its president. He improved many of the company's existing products and invented a method of bonding porcelain enamel to cast-iron fixtures. This method, which is still employed in factories today, was a key step in the history of manufacturing ... more of which later.

In the 1890s, he started experimenting with the internal combustion engine. So much so that he neglected the business, which failed. Buick and Sherwood was sold for $100,000, and in 1899, using the money from the sale of the business as capital, he opened Buick Auto-Vim and Power Company, which made gas engines for agricultural and stationary use.

In 1902 he reorganised the company to produce cars. With his machinist he developed the 'valve-in-head' overhead engine that was to make the company famous.

However, things moved disappointingly slowly and by the end of 1902 they had only produced one car; worse still, their developmental experiments had plunged them into debt. The following year, Buick borrowed $5,000 from financiers Ben and Frank Briscoe, and once more reorganised his company. The Briscoes worked out a deal with Buick that if he repaid them within four months he would regain all the stock. If not, he could lose all interest in the company. With four months coming to an end and unable to repay the loan, Buick sold his share to James Whiting of Flint Wagon Works and applied himself to the lower position of company secretary.

Whiting hired William Durant as general manager of the Buick Motor Company in 1904, and under his leadership, Buick became the cornerstone of General Motors, of which Durant became the founder. As time passed, the two clashed over various issues, including the merits of mass production versus craftsmanship, of which Buick was a staunch advocate. In 1906, Buick sold his stock to Durant for $100,000 and returned to Detroit. He made bad investments and remained in the background while the company he created evolved into one of Detroit's premier automobile manufacturers. Buick, a clever man with great ideas but little financial know-how, died in poverty.

> Buick's financial problems mean that, sadly, he was never able to afford to buy a Buick car himself.

Another Famous Car – The Rolls-Royce

Engineer and car designer Sir **Frederick Henry Royce** struck a deal with the **Rt Hon. Charles Rolls** to found the world famous Rolls-Royce Company.

The partnership started after mega-wealthy Londoner Charles Rolls studied mechanical engineering at Cambridge. He was the first undergraduate to own a car, and began racing. He set up a dealership, selling mostly foreign cars, and started to look for a supplier of reliable English cars. This led to his introduction to Huntingdonshire-born Henry Royce.

Unlike Rolls, Royce was not born into a wealthy family. When he was nine, Royce's father died and he had to work to contribute to his family's income by selling newspapers and delivering telegrams. By the time he was fifteen years old, he had completed only one year of school. He went to work as an apprentice at the Great Northern Railway company in Peterborough but had to leave after three years due to a lack of money. He then worked briefly for a tool-making company in Leeds and then for the Electric Light and Power Company

Rolls-Royce. (© Gilly Pickup)

in London. In 1884, he started a company in Manchester with his friend Ernest Claremont. He called it F. H. Royce and Company and they produced domestic electric fittings, dynamos and cranes.

In 1903, Royce bought himself a second-hand French Decauville car for the journey between his home and the factory. It was pretty useless; it was difficult to start, unreliable and overheated frequently, and it had an inefficient ignition system. He eventually became so disillusioned with the car that he decided he could do better himself. He told his colleagues that he was going to build three two-cylinder motor cars of his own design. The first of these, designed and built almost completely by Royce himself, was ready in 1904. The resultant series of two-, three-, four- and six-cylinder cars broke the mould for engineering and craftsmanship.

Launched in 1907, the Silver Ghost was a car of legendary smoothness that completed a 14,371-mile virtually non-stop run, creating the 'best car in the world' legend.

Rolls was introduced to Henry Royce by his friend Henry Edmunds, a director at Royce Ltd who drove one of the first of Royce's vehicles. Rolls preferred three- and four-cylinder cars, although Royce's two-cylinder vehicle made a big impression on him. The future co-founders of Rolls-Royce met at the Midland Hotel, Manchester, in May 1904 and reached an agreement on the foundation of a joint business, Rolls-Royce, in December 1904.

Sadly, Charles Rolls did not live long enough to see the huge success that would come their way. As the first aviator to complete a double crossing of the English Channel, he was killed in an air crash in 1910 when the tail of his Wright Flyer broke off during a flying display. He was thirty-two years of age and the first Briton to be killed in a flying accident.

During the First World War, Rolls-Royce motor cars were commissioned as ambulances, staff cars and armoured cars. Later armoured cars earned fame under the legendary Lawrence of Arabia. Even with heavily armoured bodies weighing up to four tons, they still managed to reach 50 mph.

An interesting endnote here: the **very first motor car** – ever – was down to a Brit called **Christopher Holtum**. In 1711 – yes, it's a long, long time ago – he demonstrated a horseless carriage at Covent Garden. It travelled at 5 miles an hour.

Steam Engine

When people talk about 'steam engines' they are often referring to 'steam locomotives', used for pulling railway carriages. However, steam engines were around long before Stephenson's *Rocket* completed its trials in 1829.

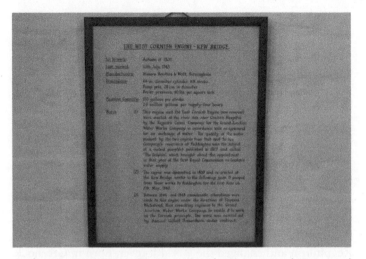

Boulton & Watt Steam Engine at London Museum of Water and Steam, Brentford. (© Mike Pickup)

The earliest records of a 'steam engine' go back to the 1690s. In 1698 **Thomas Savery**, born in Devon in 1660, patented a steam pump which could be used to raise water as a result of steam condensing and sucking up water due to atmospheric pressure. The 'engine' had a cylinder or other moving parts save for a couple of taps and, since it utilised atmospheric pressure, the maximum working height was only 30 feet above the water. In reality it was much less. The device was intended to remove water from mines and was known as the 'Miner's Friend', but because of its significant limitations only a few were built.

In 1712, **Thomas Newcomen**, born in Dartmouth in 1664, patented the first engine that harnessed steam power to produce mechanical effort. Unlike Savery's engine, which was more akin to a vacuum device, Newcomen used steam to move a cylinder up and down. This was linked by a chain to a rocking beam above it while the other side of the beam was attached to a heavy rod which was in turn attached to a pump. The rod descended due to its weight and forced water up a tube so it could be expelled at the surface. The steam engine lifted the rod on one part of the stroke and allowed it to drop on the other part, thereby creating an ongoing mechanical pumping action. One of the great benefits of this machine was that it was not limited to an operating height of 30 feet. By the late 1770s, over five hundred of these machines were in use. Various improvements were made to this device over a number of years.

Wait a minute, though – didn't Scotsman **James Watt** invent the steam engine? No, he didn't, but his improvements made it cheaper to run. In fact, his engines powered the Industrial Revolution and changed the world forever. Born in Greenock in 1736, he had a limited education because he was often ill and didn't start school until he was eleven years of age. When he went to grammar school, although he was relentlessly bullied by other pupils, he shone at maths and engineering. He became an instrument maker for the University of Glasgow when he was nineteen. Watt realised that a great deal of energy was being wasted by the repeated cooling and heating of the cylinder because cold water was added to condense the steam in the cylinder. He greatly increased the efficiency of the steam engine by introducing a separate cylinder for the steam to condense.

While this was a perfectly feasible solution, engineers of the day were more like refined blacksmiths and accurately machining a piston and cylinder proved to be very difficult. Much money was spent trying to achieve this, and one casualty was John Roebuck, owner of the Carron ironworks near Falkirk, who became bankrupt. However, the patents were acquired by Matthew Boulton, who owned the Soho Foundry near Birmingham.

Meanwhile, John Wilkinson had developed precision boring techniques for cannon making and was able to

Remaining A4 locomotives, NRM York. (© Kippa Mathews)

produce a large cylinder with a tight-fitting piston. Boulton and Watt gained access to this expertise and over the next twenty-five years developed a highly successful partnership.

There is a Boulton and Watt engine at the Brentford Steam Museum in west London. Built in 1820, the machine has a cylinder of sixty-four inches and a stroke of ninety-six inches. Its beam weighs a massive fifteen tons. Each stroke pumps 130 gallons and it is capable of just over six strokes a minute, giving it an output of 2.5 million gallons (8 million litres) every twenty-four hours.

It is worth mentioning here that the propeller is credited to James Watt too, after he first applied it to a steam engine on board a ship in 1770. Since Watt also invented the old unit of power, he had the new one named after him. Well, it was only fair really.

Thus far, steam engines were seen as pumping devices and were not concerned with movement. This change was brought about by **Richard Trevithick**. Born in Cornwall in 1771, Trevithick was a mining engineer and his early contribution to steam power was the development of a high-pressure steam engine. This meant that the boiler could be smaller and lighter so that it could carry its own weight. It also did not require the use of a condenser (thus avoiding Watt's patents). Instead it used a double-acting cylinder with a four-way valve and the beam was replaced by a crank, thereby converting linear motion into circular motion.

This enabled him to develop the steam locomotive, the first of which he built in 1801. Nicknamed the 'Puffing Devil', it was demonstrated on Christmas Eve of that year by carrying six passengers up Fore Street in Camborne, then up Camborne Hill and on to the village of Beacon.

The following year Trevithick built a high-pressure steam engine to drive a hammer, which, in 1803, under the supervision of proprietor Samuel Homfray, he mounted on wheels to create what we now know as a locomotive. He sold the patents for the locomotive to Homfray.

Homfray made a bet with another ironmaster, Richard Crawshaw, that the steam engine could haul ten tons of iron ore along the tramlines from Penydarren to Abercynon (just under 10 miles), a task normally carried out by horses. On 21 February 1804 it successfully carried ten tons of ore, five wagons and seventy men in just over four hours for an average speed of 2.4 mph.

Inspired by this development, **George Stephenson**, born in July 1781 in Northumberland, built a number of flanged-wheel locomotives, the first in 1814. This early attempt was capable of hauling thirty tons of coal up a hill at 4 mph. Traction was achieved by contact between the flanged wheel and the track, the first time this had been done.

In 1821 an Act of Parliament led to the building of the twenty-five-mile Stockton and Darlington Railway, connecting collieries in Bishop Auckland to the River Tees in Darlington. Originally planned as a horse-drawn railway, Stephenson managed to get the plans changed and later that year he and his son Robert, then only eighteen, began construction. In 1825, the first locomotive, whose name was changed to *Locomotion*, was completed and in September of that year it hauled 80 tons of coal and flour 9 miles in two hours. The first railway passenger car was attached so that dignitaries could experience the journey, thereby becoming the first passengers of a steam train.

Following this success the Liverpool and Manchester Railway was planned, and in 1829 the directors arranged to hold a competition to see who would build the locomotives. Engines could weigh no more than six tons and had to complete the sixty-mile journey. Stephenson's entry *Rocket* was driven by one of his engineers, Joseph Locke. Winning the competition made it famous, resulting in a flood of work offers from around the world, more than he could possibly cope with.

Steam engine at London Museum of Water and Steam, Brentford. (© Mike Pickup)

In 1815, George Stephenson invented a safety lamp that would not explode when used around the flammable gasses found in the coal mines.

The Glider

A pioneer of aeronautical engineering, Yorkshireman Sir **George Cayley** was the first to understand that it was not essential for a man-made flying machine to have wings that flapped like a bird's. He ignored the flapping wings and studied how birds stayed in the air as they glided around. One hundred years before the Wright Brothers, he had developed the first proper understanding of the principles of flight.

He set to work to build a model helicopter with contra-rotating propellers in 1796. In 1799 he put forward the concept of the modern aeroplane as a fixed-wing flying machine with separate systems for lift, propulsion and control.

By 1849 he had built a large gliding machine and tested the device with a (brave) ten-year-old boy aboard. The gliding machine carried the boy into the sky and fortunately returned him safely to the ground. Four years later Cayley constructed an even larger gliding machine and managed to persuade his coachman to go aboard to test the device that same year. However, this resulted in a crash landing and a rather irate coachman who said to Sir George, 'I wish to give notice. After all, I was hired to drive, not to fly.' It appears now that although the names of the boy and the coachman have been forgotten, Cayley's endeavours have not and he is regarded as perhaps the single most important aerial researcher and theoretician of his time.

Cayley could also be listed as the inventor of **seatbelts**, although American Edward J. Claghorn was granted the first patent. Cayley's 'lap belts' were designed to keep his pilots secure so that they wouldn't fall out of the glider.

Designer of the first successful glider to carry a human being aloft, Cayley discovered and identified the four aerodynamic forces of flight: weight, lift, drag and thrust, which act on any flying vehicle. Modern aeroplane design is based on those discoveries, cambered wings being an example.

Jet Engine

Frank Whittle from Coventry was the brains behind this in 1937 when he was engaged as a fighter pilot. He realised that piston-powered flight was yesterday's technology and although he was still in his early twenties and at flying school, he came up with the design for a gas turbine that could become a power source for jet thrust, patenting his 'turbo-jet' in 1930.

However, his new design was so radical that the military steered well clear of it and would not fund it. Sadly, neither would any manufacturers. Unfortunately, the Air Ministry failed to recognise the invention's potential for speed.

Fortunately, a dogged determination was the name of this young man's game. After all, hadn't it taken three attempts and a high-calorie eating regime before the RAF would even recruit the undersized Frank?

Eventually he managed to find a few private backers and work started on a prototype in Lutterworth, Leicestershire. So it was that years later, in 1941, in the darkest days of the Second World War, a test pilot powered his Gloster E28/39 off the runway at RAF Cranwell, Lincolnshire, for a seventeen-minute test flight. It was the beginning of the age of jet travel.

Sir Frank was something of a futurist, predicting the arrival of supersonic travel much earlier than faster-than-sound military jets and Concorde.

The Bouncing Bomb

Derbyshire-born **Sir Barnes Neville Wallis** was a brilliant aeronautical engineer who invented the bouncing bomb used in the 'Dambusters' raid on the Ruhr Valley.

After the outbreak of the Second World War, Churchill's government knew strategic bombing was necessary to destroy the enemy's ability to wage war. The destruction of the dams in the Ruhr, Germany's industrial heartland, had been a British ambition since 1938, but it was an impossible task without the right weapons. The Germans had designed the enormous walls of the dam to withstand the pressure of millions of tonnes of water and they believed they were impenetrable.

In 1940, Wallis, who had designed the Wellington bomber, was given the task of secretly discovering a way to destroy the walls. Although he had no funding, he came to the conclusion that a bomb bouncing off the water at a seven-degree angle would achieve a reliable bound.

He went on to devise a cylindrical bomb. Two things gave him inspiration. One was the 'ducks and drakes' game of skimming stones across water, which he used to play with his family, and the other was naval gunners of old, who knew how a cannonball's range could be extended by skimming it over water.

Although the authorities and the government had serious misgivings, Wallis persevered to create the bouncing bomb in

March 1942. Sir Arthur Harris, Chief of Bomber Command, initially described the idea as 'tripe of the wildest description' and claimed that 'there is not the smallest chance of its working'. He said it was 'just about the maddest proposition as a weapon we have yet to come across', adding, 'I am prepared to bet my shirt that the bomb will not work when we have got it.'

Wallis ignored it all. Robert Owen, official historian of the 617 Association, said, 'He was a forward thinker, he thought outside the box and you can imagine that anyone who came across him thought that his ideas were fanciful and crazy. He was a remarkable engineer, but it was all to his credit that he also had perseverance and self-belief. Everyone around him said, "You can't make four tonnes of metal bounce" and he said, "Yes, I can."'

With just eleven weeks to go until the dams were full, Wallis did not have a full-scale drawing of the bouncing bomb, known as the upkeep mine. As work began he was one of very few people who knew the purpose of what was known as Operation Chastise. The men building the components were kept in the dark.

As a teenager, Ron Hyde was working in one of the hangars. He remembered, 'During these tests and modifications Dr Barnes Wallis never seemed to leave the area or sleep. He wore his glasses at the end of his nose and was oblivious to those around him except when asked a question or if he had a specific requirement. He was always on the move, making copious notes and working out calculations on his slide rule. He never seemed to stop.'

The bombs and the modified planes were completed just in time. As Wallis had known all the time, the sceptics were to be proved wrong. Following a successful test run on a disused dam in Wales, the Dambusters raid was given the go ahead in 1943.

When Bomber Command took to the air Wallis was in the Grantham RAF headquarters anxiously waiting to find out if his design had worked. It had.

Ten minutes before the target was reached an electronic motor was used to spin the bomb to 500 rpm, which Wallis had calculated would stabilise its flight. The thirty-tonne altered Lancaster had to approach the dam at the right height of 60 feet and the correct speed of 230 mph to deliver the precision bomb. Spotlights were used to determine the aircraft's height and at 60 feet they intersected and created a figure of eight on the water. At 1,276 feet from the dam, the men were to release the bomb. It bounced across the water, over the anti-torpedo defences, and sank at the wall where it was programmed to detonate at 30 feet. When it sank, the backspin caused the bomb to stay close to the face of the dam, focusing the force of the explosion against the wall.

The raid delivered a humiliating blow to Hitler's Germany. Both the Möhne and Eder dams were breached, damaging German factories and disrupting hydroelectric power. A grateful British nation gave thanks.

Dundee University lecturer Dr Iain Murray, author of *Bouncing-Bomb Man: The Science of Sir Barnes Wallis*, believes that this logical scientific mind was the true genius of the inventor. 'Wallis was able to make significant novel contributions to his field because he had been taught solid scientific principles at school, he knew a lot about many different disciplines, and he knew how to find out more about other areas as and when needed. He could bring all of this knowledge to bear on the problem at hand.'

After the war, Wallis designed the 'swing wing', which was tested on model planes. His daughter, Mrs Stopes-Roe, said, 'The death rate after the Dams raid upset him so profoundly for the rest of his life. He gave away his inventor's award to education and that's why he would never allow a test pilot to test fly the "swing wing" as he said he would never endanger another man's life. If he was here today his thoughts would be with the crews. He saw the government's £10,000 award as blood money for the crew.' Remembering how hard her father worked, she added, 'He tried to put into practice all that he believed and felt was the right thing to do.'

His other projects included a stratospheric chamber for aircraft testing, bridges, a radio telescope and a jet-powered submarine.

As is sometimes the case, a group of people can work concurrently on the same idea. This can often cause confusion over who gets credit for coming up with the idea. The light bulb, the cycle, calculus, pneumatic tyres, the microchip and photography are all examples of parallel inventions...

Military Tank

Likewise, no single individual was responsible for the development of the tank, and its original design goes back to the eighteenth century. Over time, gradual technological developments brought the development of the tank as we know it closer, until its eventual form was unveiled by the British Army – or rather, Navy, since its initial deployment in the First World War was overseen by the Royal Navy.

Ernest Swinton was an Army officer and Britain's official war correspondent. He suggested in 1914 the crawler tractors used to pull artillery on the Western Front could be used as offensive weapons, with their capability to climb a five-foot obstacle, span a five-foot trench, resist small-arms fire and travel at 4 mph. Both Swinton and the Secretary of the Committee for Imperial Defence, Maurice Hankey, were enthusiastic about what they believed to be the enormous potential of the tank. While Hankey produced the first official memo about the tank on 26 December 1914, Swinton organised a demonstration of the Killen-Strait vehicle to senior politicians in June 1915, almost a year after the war was underway.

Colonel Swinton, backed by Hankey, urged Churchill to sponsor the establishment of the Landships Committee to investigate the potential of constructing what was a new military weapon. The committee's name originated from the fact that, at least in the beginning, the tank was seen as an extension of seagoing warships, for this reason being known as a landship.

So the first ever tank was built at a Lincoln engineering company, **William Foster & Co. Ltd**, in 1916. The company was approached by the Admiralty Landship Committee to design a prototype. This was top-secret work and was carried out in a room at the city's White Hart Hotel by Foster's MD, William Tritton, his chief draughtsman, William Rigby, and Major Walter Wilson of the War Cabinet.

The first design was deliberately falsely described as a water carrier for Mesopotamia to hide its real purpose. Workers had to refer to them as 'water tanks' or just 'tanks' and the name stuck.

On 22 September 1915, Tritton sent the following telegram to the Admiralty:

> New arrival by tritton out of pressed plate STOP
> Light in weight but very strong STOP
> All doing well
> Thank you STOP
> Proud parents

Today's tanks are still recognisable when compared with the original design made in Lincoln.

Propeller-Driven Steamship

Isambard Kingdom Brunel, born in Portsmouth, was a renowned civil engineer credited with, among other things, constructing a network of tunnels, bridges and viaducts for the Great Western Railway and the design of its terminus, London's Paddington station.

He also designed a number of bridges, most notably the Bristol Suspension Bridge, which, at the time, had the longest span of any bridge worldwide. However, it was not completed until five years after his death.

He envisaged extending the Great Western Railway from Bristol to New York so passengers could book a single ticket to cover the journey from London to New York. To achieve this he designed the *Great Western*, some 235 feet long and at that time the largest steamship in existence. The ship first sailed across the Atlantic in 1838.

His great invention was, however, the propeller-driven steamship. Up to that time most ships, even those crossing the Atlantic, were driven by paddles.

Following successful trials of a propeller on the tug *Archimedes*, his design for his new ship, the 332-foot *Great Britain*, incorporated a propeller comprising six blades. The ship was made of metal rather than wood and in 1845 became the first propeller-driven iron ship to cross the Atlantic, carrying up to 250 first- and second-class passengers.

The ship was later run aground but has since been completely restored and is on display in Bristol, where it is open to the public.

SS *Great Britain*. (© mattbuck under Creative Commons 2.0)

Cruise Holidays

The rivers and seas of Europe have always been essential to life and trade but, until the mid 1800s, boats and ships were only concerned with moving cargo or hauling in the day's catch, not with transporting passengers.

Shetland sailor **Arthur Anderson,** a man of remarkable negotiating skills, changed that idea in 1835 when he proposed the idea of sailing for pleasure as a passenger in an ocean-going vessel. He launched a newspaper called the *Shetland Journal* and wrote a spoof advertisement for 'cruises' around the Shetland Islands to Iceland and the Faroes. His vision was to provide passenger services from Scotland to Iceland in the summer months and from Scotland to the Iberian Peninsula (Spain and Portugal) in winter.

Just two years later his dream moved closer to reality, when he and Brodie McGhie Wilcox co-founded the General Steam Navigation Company, later the Peninsular Steam Navigation Company and now known as P&O, a major operator of passenger liners. On 1 September 1837, Anderson and Wilcox were awarded a contract from the British government to carry mail from London to the Iberian Peninsula. They also offered voyages known as 'excursions' when passengers from Britain travelled with the mail returning home on other P&O mail voyages.

It was not long before P&O's routes expanded east, and in 1840 they were awarded a new contract to extend their service to the Egyptian port of Alexandria via Malta. The new contract required that the voyage to Alexandria be complete in fifteen days. The first vessel to operate on this service was the newly built 1,787-ton paddle-wheel steamship *Oriental*, reflecting the company's arrival in the east. The company name was now the Peninsular & Oriental Steam Navigation Company.

P&O Cruises. (© Gilly Pickup)

In 1842 routes expanded to India, and in 1845 Penang and Singapore were added. Seven years later, the SS *Chusan* began sailing to Australia. Though P&O's main focus was mail delivery, it soon became clear to both men that there was more to a life at sea than just getting from A to B. On 14 March 1843, P&O placed a historic, pioneering advertisement in the *London Times* for a round voyage in the 782-ton paddle steamer *Tagus*. The advertisement read, 'Steam to Constantinople, calling at Gibraltar, Malta, Athens, Syria, Smyrna, Mytilene and the Dardenelles.' From that advertisement, P&O continued to develop the popular 'classic' voyages.

So came the birth of leisure cruising, with the first leisure cruise departing London bound for the Mediterranean in 1844. That same year, novelist William Makepeace Thackeray sailed around the Mediterranean as a guest of P&O and on return wrote his book *From Cornhill to Cairo*.

Foreign travel became fashionable among the newly wealthy of the Industrial Revolution, although even then sailing for pleasure did not really become popular until the twentieth century. Other shipping lines that started out by carrying mail across the Atlantic also saw the potential and began to offer passenger services.

More ships began to consider the comfort of passengers, and in 1840 the *Britannia*, the first ship to sail under the Cunard Line name, reportedly took a cow on board to supply fresh milk on a transatlantic crossing.

During the first half of the twentieth century, liners were built to serve passengers travelling between Europe and North America. Bigger and better ships which took on the characteristics of elegant, floating hotels were built and competed to make the fastest crossing of the Atlantic. They raced for the Blue Riband trophy, awarded to the fastest transatlantic crossing. At the start of the Second World War, the necessity of transporting enormous numbers of troops and personnel around the world required that all British registered passenger and cargo ships were put into service as troop carriers. However, from 1945 to 1972 it was from the decks of P&O ships that over 1 million United Kingdom migrants caught their first glimpse of their new Australian homeland. Known as Ten Pound Poms, their mass arrival was a scheme devised by the Australian and British governments to help populate Australia.

Interest in transatlantic cruising surged between the wars and again after the Second World War, but a decline began in the late 1950s and during the 1960s the European cruise industry slowly refocused on sailing the coasts and rivers of the Continent. By the 1970s the development of the jet engine and long-haul passenger aircraft saw a dramatic reduction in passengers using cruise ships. Now, in the twenty-first

The naming ceremony of P&O *Britannia* by Her Majesty the Queen. (© Mike Pickup)

century, more and more cruise ships are being built. P&O's largest ship built exclusively for Britain, *Britannia*, an iconic modern classic, entered service in spring 2015 and was named by Her Majesty Queen Elizabeth II.

With a length of 1,082 feet, she carries 3,647 passengers and at 141,000 tons is 25,000 tons larger than any other ship in the P&O Cruises fleet. *Britannia* also displays on her bow P&O Cruises' bold new livery, with the longest version of the Union Flag anywhere in the world at 308 feet.

There have been two previous ships named *Britannia* connected to the company. The first entered service in 1835 while the second, which entered service in 1887, was one of four ships ordered to mark the Golden Jubilee of both Queen Victoria and P&O itself. This Golden Jubilee-class ship carried 250 first-class passengers and 160 second-class passengers and had a cargo capacity of approximately 4,000 tons. In 1888 the young Winston Churchill sailed on this *Britannia* with his hussar regiment to Bombay and then went

Britannia. (© P&O Cruises)

on to fight on the North West Frontier. Whilst the name *Britannia* has great historical resonance with P&O Cruises, the newest member of the fleet celebrates the forward-looking Britain of today and the future.

What would Arthur have thought of it all? One can only imagine.

As a child, Arthur Anderson worked on the beaches of the Shetland Islands, preparing fish for market. The Crown tried to press-gang him for service but he was allowed to join the Navy in 1808 and was discharged ten years later.

The word 'posh' originates from this period. In these days before air conditioning, Britons travelling on a vessel to India would favour a cabin on the shaded side of the ship, away from the glare and heat of the sun. Thus when travelling from the UK to India, a north-facing port cabin cost more than a south-facing starboard one. The opposite applied on the return journey. So only the wealthiest could book a cabin that was PORT OUT, STARBOARD HOME. This was shortened to 'posh'.

The Package Holiday

A package holiday is primarily a package of travel and accommodation purchased as one item and put together by a tour operator, although more recently the accommodation element may be replaced by another destination offering, such as car hire. **Thomas Cook**, a cabinetmaker from Market Harborough, is credited with inventing the package holiday although his first foray into this area appears not to have included the provision of accommodation. However, what it did prove to the young Cook was the concept of economies of scale.

On 9 June 1841, thirty-two-year-old Thomas walked from his home to the nearby town of Leicester to attend a temperance meeting. A religious man, he believed that many social problems were related to drink. On the journey he realised that more people would attend these meetings if they were easier to reach, so he arranged with Midland Railway Company to charter a train and a month later some 500 passengers paid the princely sum of one shilling for the twenty-four-mile round trip from Leicester to Loughborough. Thus was born the idea that quickly developed into early package tours, although these initial trips were philanthropic rather than motivated by the desire to make a profit.

Cook's first commercial venture was in 1845, when he organised a trip to Liverpool. As well as offering low-priced fares, he investigated the route and produced a sixty-page handbook of the journey, the forerunner of today's holiday brochure.

In 1851, he was responsible for organising 150,000 visits to the Great Exhibition in London and in 1855 he planned a tour to the International Exhibition in Paris. Unable to get cooperation from cross-channel ferry operators, he was forced to use the crossing from Harwich to Antwerp. However, this presented him with the opportunity to create a tour encompassing Brussels, Cologne, the Rhine, Heidelberg, Baden-Baden, Strasbourg and Paris and during the summer of 1855, using this route, Cook escorted his first tourists to Europe.

He continued to expand his operation across Europe, encompassing Switzerland and Italy in his itineraries. As part of this expansion, he needed to negotiate with hoteliers to provide good food and accommodation for his guests. Keen to get his business, they were happy to work with him and as a result Cook made two further inventions. One was the hotel coupon, which his guests used to pay for accommodation and meals and which later became the accommodation voucher. The other was the circular note, first issued in 1874, which guests could use to obtain cash in a local currency, the first example of the traveller's cheque.

By 1872, Cook was ready to take on the world and embarked on his biggest tour, a 222-day tour to Egypt via the USA, Japan, China, Singapore and India. The journey covered more than 25,000 miles and cost 200 guineas.

The business, in the hands of Thomas Cook's descendants, continued to flourish but Thomas Cook & Son was sold, in 1928, to the *Compagnie Internationale des Wagons-Lits*, owners of the Orient Express and other luxury sleepers. Shortly after the outbreak of the Second World War, the company's headquarters in Paris were seized by occupying forces and Thomas Cook & Son's UK assets were seized by the British government. After the war, the financial affairs of the company were resolved and in 1948 the company became a state-owned part of British Railways.

Meanwhile, the package holiday as we know it today began to take shape in the mind of Vladimir Raiz. Born in London in 1922, he attended Mill Hill School and started work as a journalist in 1942. While on holiday in Corsica in 1949, he was asked if he could encourage more British tourists. He calculated that he could charter a plane and arrange two weeks' accommodation and food for less than £35. However, in those highly regulated days he needed permission from the Ministry for Civil Aviation. Since the cost of the two-week trip was probably lower than the cost of just a normal flight, the Ministry approved the idea on the basis that it was only for students and teachers.

On 29 May 1950, the flight, operated by his newly formed company Horizon, left Gatwick bearing eleven 'teachers' (in fact fare-paying guests) and twenty-one friends. Other tours were launched to Palma in 1952, Lourdes in 1953 and Sardinia and the Costa Brava in 1954. In the same year, amendments made to the Convention on International Civil Aviation allowed mass tourism using chartered planes, which opened up the charter flight holiday market as we know it today.

The market continued to grow rapidly as UK holidaymakers enjoyed sunshine holidays with flights, transfers and hotels with their own bathrooms, often at lower prices than hotel holidays in the UK, where bathing facilities were usually shared and good weather could not be guaranteed.

One of the major participants in this was Court Line, originally a shipping company. Court Line Aviation came into being on 1 January 1970 when it took delivery of the first of several BAC One-Eleven 500s. Court Line was responsible for two memorable innovations. First it painted its aircraft in a range of pastel shades, and Mary Quant designed matching trendy uniforms for the aircrew. Flying was a serious business in those days but the Court Line image reflected a fun approach, making the flight an enjoyable part of the holiday.

Court Line also introduced the concept of seat-back catering. In the back of the seat in front of them, passengers were faced with two compartments, each containing a cold meal. On the outbound journey one compartment was locked in order to prevent the outbound passengers eating the meal intended for the inbound passengers, although this lock was not entirely foolproof! Since no meals were prepared on board, this arrangement allowed for a smaller galley and therefore more seats. It also meant that aircrew did not have to handle food trays in flight, both resulting in considerable savings.

Package holidays continued to rise in popularity for many years but the increased presence of low-cost airlines tempted a growing number of British holidaymakers to adopt a DIY approach, booking their flights, accommodation and other items independently. However, the failure of several companies in the last few years has led to the resurgence of the package tour. This is largely due to the protection afforded by the Air Tour Organisers Licence (ATOL). This scheme, operated by the Civil Aviation Authority, applies to any holiday purchased from an ATOL holder that includes a flight plus one other element, usually hotel accommodation, cruise or car hire. The scheme protects customers against any failure of the tour operator or any of its suppliers.

The package tour has come a very long way since that first train journey in 1841. Thanks to Thomas Cook for getting the whole thing started.

Package holidays remain popular, with 46 per cent of holidaymakers opting for an overseas package holiday in 2014 (ABTA survey).

And to complete this section, here are some more Great British Inventions in the Movement category ...

Electric Motor

Invented in 1821 by Londoner **Michael Faraday**. He was working at the Royal Institution when he demonstrated electromagnetic rotation for the first time. A free-hanging wire was dipped into a pool of mercury that had a fixed magnet in it. When an electric current was passed through the wire, it rotated around the magnet, the electricity producing a magnetic field around the wire, which interacted with the magnet in the mercury. And there you have it – the world's first electric motor.

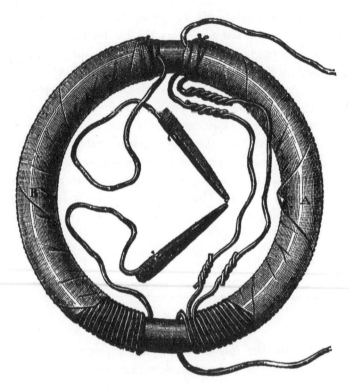

Faradays transformer by Friedrich Uppenborn (1859–1907).

Speedometer

Scottish electrical engineer **Sir George Keith Elphinstone** specialised in electrical instruments for use on the road. He developed speed-recording equipment for trains and then a similar device for cars in the early 1900s. Other inventors developed their own variations, but Elphinstone was the first. He died in 1941.

The Hovercraft

Christopher Cockerell was a brilliant engineer from Cambridge who in 1956 invented the hovercraft. The vehicle, also known as an air-cushion vehicle or ACV, can travel over land, water, wetlands, swamps, mud and ice. Originally, in 1955, he had experimented with the idea by using an empty KiteKat cat food tin inside a coffee tin, an industrial air blower and a pair of kitchen scales. He was trying to discover if it was possible to produce a cushion of air between the bottom of the tins and the surface of the scales. Once he found that this was possible he decided to experiment with more sophisticated models. And so the 'hovercraft' came into being. Hovercraft have been put to a variety of uses, both military and civilian – but there's more, because not only did he invent the vehicle, he also invented the name! The first hovercraft successfully crossed the channel on 1 June 1959.

Jet Propulsion

Sir Isaac Newton forecast that one day people would travel at 50 mph. In 1680 a man called Gravesande designed a car-type vehicle that would be powered by Newton's third law of motion: 'To every action there is an equal and opposite reaction.' A boiler sent out a jet of steam that pushed the car along. Everyone on the road behind the jet engine would have been scalded, but perhaps that is a small price to pay for progress.

Torpedo

In 1866 engineer **Robert Whitehead** from Bolton designed a torpedo to be launched from a ship in an underwater tube, powered by compressed air and with an internal mechanism that adjusted itself to stay at a constant depth. The first ship sunk by his invention was Turkish steamer *Intibah* in 1878, after being hit by a torpedo launched from a Russian warship.

Disc Brakes

Invented in 1902 by Londoner **Frederick Lanchester**, disc brakes employ brake pads that squeeze each side of the rotor turning a wheel. They were quicker to cool down and to dry out than the drum brakes used in most cars at the time. Sadly, they didn't catch on until the 1950s, after Lanchester's death, but nowadays almost all cars use his invention.

3

Technology

Telephone

Who's calling please? Edinburgh-born **Alexander Graham Bell** invented the 'electrical speech machine', which we know as the telephone. He first became interested in the science of sound because both his mother and wife were deaf.

His father was a teacher of elocution and speech correction, pioneering a method of teaching the deaf called 'Visible Speech'. Mainly self-taught and educated at home, only spending two years at Edinburgh Royal High School, young Alexander had a thirst for knowledge and a fascination with modes of communication. After both of his brothers died of tuberculosis, he emigrated to Canada in 1870 with the rest of his family. By that time Bell had already followed in his father's footsteps as a teacher of Visible Speech.

On 14 February 1876, Bell and an American electrical engineer named Elisha Gray, both filed patents with the US Patent Office covering the transmission of sounds telegraphically. Who got there first is debatable but Bell's patent, listed as 'Improvement to Telegraphy', was awarded. This meant he could get funding and was able to hire an assistant, Thomas Watson, who helped him to create a machine for transmitting sound electronically. They worked until they were exhausted, but it was worth it when Alexander spoke the first words over the telephone on 10 March 1876,

'Mr Watson, come here, I want to see you.' There was a problem, though, in that the first telephone didn't have a bell, so the caller had to tap the phone with a hammer to let the receiver know a call was being sent to them. Thomas Watson was the brains behind the invention of the bell.

Bell insisted on 'ahoy!' as the correct way to answer the telephone. It was trounced by 'hello', which became the standard telephone greeting.

The telephone made Bell hugely successful and he was able to leave the board of the Bell Telephone Company at the age of thirty-two in 1879. Now he could pursue other ideas. He bought a chunk of land on Cape Breton Island in Nova Scotia in 1885, established laboratories and went on to continue his research and his long line of inventions.

He went on to invent the gramophone and the wireless photophone. The latter transmitted sound via a beam of light and was the precursor of modern fibreoptics. He also implemented new techniques for teaching the deaf to speak.

In 1898 he became the president of the National Geographic Society, believing that photographs of the wider world would bring an understanding of geography to the many people who would never travel for one reason or another.

He didn't stop there, though. His interest in aeronautics and kites led him to build a rather strange looking man-carrying kite. The Cygnet, one of his kite designs, went on to achieve the first controlled manned flight in 1907. Something of a record-breaker by now, Bell also invented a hydrofoil. It set the world water speed record in 1919. Besides all of that, he also invented the first metal detector.

When he died in 1922, he had collected eighteen patents and twelve others that he shared. Not bad for a man with such limited formal schooling.

He didn't have the middle name 'Graham' until he turned eleven, when his father gave it to him as a birthday present. Alexander had asked if he could have a middle name like his two brothers.

Programmable Computer

Despite working in the mechanical rather than the electronic age, Londoner **Charles Babbage** is regarded as the father of the programmable computer.

As a sickly child, and a banker's son Babbage was mainly educated at home before he left to attend Cambridge University to study mathematics. He was keen to find a way to eliminate errors in complicated calculations since these required the use of printed mathematical tables and were apt to be inaccurate.

Babbage started work on his 'Difference Engine' in the 1820s, a machine which could perform mathematical calculations. A six-wheeled model was initially constructed and demonstrated to a number of audiences. If it had been completed, it would have had 25,000 parts, stood 8 feet long, 7 feet high and 3 feet deep. Rather unwieldy.

He then developed plans for a bigger and better machine, known as 'Difference Engine 2'. A point worth noting is that in 1985, when the Science Museum built his Difference Engine 2 from original drawings, it contained 4,000 moving parts, weighed 2.6 tonnes and worked perfectly.

This was followed by a much more complex affair in 1834, the Analytical Engine. A major intellectual feat of the age, it possessed all the features of a basic modern computer, and it is on this that Babbage's fame as a computer pioneer now largely rests.

It was intended to be able to perform any arithmetical

calculation using punched cards that would deliver instructions, as well as a memory unit to store numbers and many other fundamental components of today's computers. His inventions were heavy, cumbersome, incredibly complex and, partly because of this, never completed. Had Babbage's Analytical Machine received the financial backing and development it needed, who knows, it might have kick-started the computer age a century sooner.

> British mathematician Ada Lovelace, more on whom follows, completed a programme for the Analytical Engine but neither it, nor Difference Engine 2, were finished in Babbage's lifetime.

Computer Programming

Ada Lovelace, the daughter of Annabella Millbanke and Lord Byron, was born in London. She was the poet's only

Watercolour portrait of Ada King, Countess of Lovelace, 1840. (Science & Society Picture Library)

legitimate child. Not that he ever knew his daughter; his wife fled in the night with her when she was only weeks old. Theirs was a short and often brutal marriage, with Lord Byron having multiple affairs.

Annabella was desperate to make sure her daughter would not become a reprobate like her father and encouraged her daughter to study the sciences and mathematics, although at that time these were not considered ideal subjects for girls of Ada's social class.

However, Ada loved studying – she also became fluent in French – and was a brilliant child who, when a youngster, intended to invent a flying machine. This future 'Lady Fairy', as inventor Charles Babbage affectionately called her, went about the project methodically. Her first step, in 1828, was to construct wings. She considered various materials for the wings – paper, oilsilk, wires and feathers – and then decided to write a book, *Flyology*, illustrating some of her findings.

When she grew up she married William Lovelace, though she was very close to Charles Babbage and his unfinished Analytical Machine caught her attention. She expanded on the original writings and described an algorithm for the Analytical Engine to compute an established sequence of numbers, making her the first ever computer programmer. Unfortunately the Analytical Engine was never completed, so there was no way to test her theory, though it did assure her place in computing history.

Ada saw the machine's real potential. In 1843, she published her translation of Italian mathematician Luigi Menabrea's article about the Analytical Machine. With the article, she appended a set of notes. She published under the initials 'A. A. L' to hide her identity since at the time women were often not accepted as intellectual equals.

Explaining the Analytical Engine's function was a difficult task, as even other scientists did not really grasp the concept and the British establishment had no interest whatsoever in it. Ada's notes even explained how the engine differed from the original Difference Engine. Whether or not it was her intention, Ada laid the groundwork for the development of the computer as we know it today.

Babbage was impressed by Lovelace's intellect and analytical skills, calling her 'the Enchantress of Numbers'. However, because of his refusal to accept help from a largely self-educated woman, we will never know what might have been.

The World Wide Web

Can you imagine a world before Twitter, Google, Facebook ... a world before the World Wide Web?

'This is for everyone.' Four words, in huge, animated letters, swirled around the Olympic Stadium during the opening ceremony for London 2012. The message was activated by a computer, by a man in the middle of the stadium. He was Sir Tim Berners-Lee, whose World Wide Web has been compared in importance to Gutenberg's printing press by *Time* magazine ...

... But let's start at the beginning. In the Cold War era of the 1960s, the USA and Russia were involved in an arms race, the winner of which, some believed, would gain global dominance. In the US, the Advanced Research Projects Agency (ARPA) was leading much of the work and they realised the benefit of creating a network which would allow geographically widespread research institutions to share experiences and findings.

Computer networking was also beginning to be used in the commercial environment within large corporate buildings using direct physical connections, but there was no standard system. Commercial networks used a variety of proprietary protocols (languages) to allow computers to talk to each other, and these were not compatible with each other. Data communication was in its infancy and not very reliable, so ARPA created a network consisting of computer switches, or hubs, that would allow data sent from one part of the US to another and to use different routes to reach their destination in order to improve reliability. This was the first ever creation of the packet-switched network, known as ARPANET. It also allowed proprietary network protocols to be translated into ARPANET's protocol by the sender of the data and translated into the protocol of the recipient's network so it could be read. Some years later ARPANET was effectively de militarised and made available to commercial users. As a network which sat between other networks it was called the Internet, and its protocol became known as Internet Protocol (IP).

However, the use of this network was still limited primarily to the academic and research markets and it was, it has to be said, a bit of a hit-and-miss affair. Looking for information required logging on to a remote computer, searching the files and downloading documents whose titles suggested they might be relevant to an enquiry. Data speeds were slow, downloads could take many minutes and documents could not be read until they had been downloaded, often resulting in disappointment for the recipient.

Enter London-born physicist and computer scientist **Tim Berners-Lee**, who studied Physics at Oxford University. He was working at the European Organisation for Nuclear Research (CERN) on the Swiss–French border in the late 1980s when he sought to find a better way for his colleagues to link up. He hit on the idea that combining a method of encoding links into words (hypertext markup language or HTML) together with the power of the internet could greatly speed up this process. He found no commercial organisations that shared his view.

By the end of 1990 Berners-Lee had built all the elements necessary to deliver his new concept, including the first browser and the first website (http://info.cern.ch) and the rest, as they say, is history.

The technology was also made freely available and without royalty. Had he not done so he would almost certainly be one of the world's richest people. However, because it was freely available, adoption of the technology was very rapid and today it is estimated there are over 2 billion users of Berners-Lee's invention. Where would we be without it today?

The story goes that in looking for a name for this new invention, he turned down 'The Information Mine' – TIM for short, matching his name – and modestly called it the World Wide Web.

Traffic Lights

We have **John Peake Knight,** a Nottingham engineer, to thank for this essential piece of traffic management on our roads today. Strictly speaking Knight wasn't really an inventor, though he could always spot a good idea and make use of it. In the mid-1860s, Knight was a railway manager specialising in designing signalling systems for Britain's ever-growing railway network. It seemed obvious to him that these should be adapted for use on the road, so he approached the Commissioner of the Metropolitan Police with the idea for using the basis of a railway signalling system on the streets of London. Knight's proposal was to use a semaphore system, which determined whether a train could pass or not with an arm that goes up and down to indicate whether to stop or go. At night, when it was too dark to see the arms, the traffic signal would switch to using red and green gas lamps.

At the time there were no cars on the road, but there was growing concern at the number of horse-drawn carriages and the increasing danger to pedestrians. Knight decided to treat main roads and side roads in the same way as main and branch railway lines. The semaphore method would be used during the day and at night gas-powered red and green lights would be operated.

It was three years before the plan was implemented, but on 9 December 1868 the world's first traffic lights were installed in London at the junction of Great George Street and Bridge Street in Westminster, close to Westminster Bridge. People came from all around to see and marvel at this new idea, and it was an immediate success. Within a short time more signals began to appear in the centre of London. They were operated by policemen who had to stand next to them all day. But then, disaster struck one evening when a leaky gas mains resulted in one of the traffic lights exploding in the face of the policeman who was operating it. He was badly burnt.

The project, which had been praised so enthusiastically, was immediately dropped. It would be another forty years before traffic lights reappeared, but it wasn't in the UK, it was in America. The first three-colour traffic lights in Britain were installed in London at Piccadilly in 1925 and they were operated by policemen. Automatic signals, working on a time interval, were installed in Wolverhampton in 1926. However, traffic lights only became a common London sight in 1929, when the first electric signals were introduced.

In 1932, the first vehicle-actuated signals in Britain appeared on the junction between Gracechurch Street and Cornhill. It was a real quirk of fate that these too were destroyed by a gas explosion.

Knight is also credited with being one of the first to introduce emergency brake cords in trains.

The Television

Just imagine, without him there would be no *Eastenders*, no *Britain's Got Talent* or *I'm a Celebrity, Get Me Out of Here*. But still, we can't condemn him entirely for giving us the television. However, at the start, **John Logie Baird** did have problems with his brainwave. One of his initial experiments in his Hastings laboratory, when he used a transmitter and receiver created from biscuit tins, penny lenses and string, led to a small explosion which in turn led to his landlord evicting him. He was also responsible for causing a blackout across Rutherglen when he attempted to produce artificial diamonds by passing an enormous current through a stick of graphite.

Scottish engineer Baird invented the television in 1925, although it is hard to credit just one person with the invention. What is indisputable is that Baird was the first to transmit moving pictures in October 1925, but his mechanical system ultimately failed, with a rival being developed at the same time able to produce a visibly superior picture. Baird, it was said at the time, was 'doomed to be the man who sows the seed but does not reap the harvest'.

Born in 1888 in Helensburgh, Baird, who was plagued by illness, first demonstrated his 'televisor', a televised moving image of a ventriloquist's dummy, in his attic laboratory at 22 Frith Street, Soho, in 1926. In 1927 he broadcast the first long-range television transmission, which covered a distance of 438 miles from London to Glasgow. The first public demonstration of his televisor was held at Selfridges

1950s room with television. (© Gilly Pickup)

a year later, but it was not until 1928 that Baird showed off a proper working version. The same year, he also began experimenting with colour TV and on 3 July transmitted the world's first colour televisual images. He set up the Baird Television Development Company, which made the first televisual transatlantic crossing between London and New York. Then, in 1931, he broadcast the Epsom Derby live for the first time, which was also the British Broadcasting Corporation's first television broadcast.

From then till 1932, BBC transmitters broadcast TV programmes using the thirty-line Baird system – before later switching to a rival electronic system. In 1932 he made the first television broadcast between London and Glasgow.

In 1936 an early BBC transmitter, on a hilltop at Alexandra Palace in London, began beaming pictures created on his mechanical, 240-line system. EMI-Marconi's rival electronic system, using 405 lines and trialled simultaneously, ultimately won the day but Baird was the pioneer and went on to invent colour, stereoscopic and big-screen television. By the time the Coronation of Queen Elizabeth II was televised in 1953, the 'small screen' was the must-have household gadget.

To see some of Baird's original apparatus in the National Media Museum in Bradford is to return to the exciting early days of grey, flickering 'pictures by wireless'.

And there is an addendum to this tale. Not only was Baird the first person to transmit live television images, but during the First World War he invented something else. Any idea what that might have been? Well, the trenches were awful places to be if you were a soldier – wet, muddy, cold, uncomfortable. The soldiers weren't able to change their clothes or socks very often, and this led to what was known as 'trench foot'. Untreated, it led to all manner of awful ills, including amputation. Baird devised an 'undersock' to keep the soldiers' feet healthy. Marketing at the time included testimonials from some of the soldiers, though it isn't clear if these were absolutely genuine or not. A Corporal H. G. Roberts is reported as saying, 'I find the Baird Undersocks keep my feet in splendid condition out here in France. Foot trouble is one of our worst enemies, but, thanks to the Baird Undersock, mine are in the "pink", and I think they should be supplied to all soldiers.'

Baird had tried to enlist in the war but was deemed unfit for military service because of the illnesses he had suffered since he was a boy.

His daughter Diana explained where he got the idea for the undersock. 'He had cold feet,' she said. 'His circulation was absolutely awful. He was always cold. I remember him wearing big, thick, heavy overcoats and he was still cold. He just could not get warm.'

Stewart Noble of Helensburgh Heritage Trust, said, 'He discovered that if he put newspaper inside his shoes that would absorb the moisture that always comes from the feet. He came up with this idea of the Baird undersock, which was basically a sock sprinkled with borax. This would absorb the moisture that the foot gives off.'

To advertise the sock Baird first put an advert in the *People's Friend*, but it was a waste of time as there was little response. He put his thinking cap on and decided to get people's attention by making a plywood mock-up of a tank covered with adverts for the sock. He then paid someone to push it around Glasgow. He also employed sandwich-board women to walk around advertising the socks. 'His innovation was having ladies do it instead of men,' says his daughter.

The product was a success and it meant he was making enough money to leave his job as assistant mains engineer with the Clyde Valley Electric Power Company, which he hated. In his memoirs he described the job as 'sordid miserable work, punctuated by repeated colds and influenza'. Baird was unlikely ever to have progressed in his work due to his constant ill health. Mr Noble said that in twelve months' work on the Baird undersock he had earned as much as he would have in twelve years with the electricity supply company.

In his marketing he skilfully mixed the promise of healthier feet with testimonials from soldiers in the British Expeditionary Force in France. One from a 2nd Lieutenant G. H. stated, 'They are the very things required out here. Woollen socks get sticky and "clammy" and we can't get them washed. The Undersocks keep the feet and the ordinary socks fresh for weeks.'

Ms Richardson says, 'It was doing very well. It was booming but it was a one-man business. When he disappeared for six weeks the business disappeared too.'

According to Mr Noble, 'He was once again hit with one of his very bad colds so he just closed it down at that point and discovered that at the end of the day he had got something like £1,600 in the bank.'

Baird saw this as an opportunity to move to a warmer climate and sailed off to Trinidad, where he experimented with jam making. This venture failed and he returned to Britain, but his trip to the Caribbean may have sown the seeds for his most famous invention. Mr Noble says, 'There are tales of this fair-haired white man in a shack in the jungle with bright flashing lights. Whether or not that was him doing television experimentation I don't think anyone really knows but there is a possibility he was trying something out there.'

Once, when accused of being absent-minded, Logie Baird tapped his head and said, 'I'm present-minded in here, I'm only absent-minded on the outside.'

Electric Light

Wait a minute though – wasn't it Thomas Edison who got there first? No, he didn't actually. Englishman Sir **Humphry Davy**, who experimented with electricity and came up with an electric battery, invented the first incandescent electric light in 1800. After connecting wires from his battery to a piece of carbon, electricity raced between the carbon pieces, producing an intense, hot, short-lived light. Davy didn't stop at that, though; he also invented a miner's safety helmet and a process to desalinate sea water. And that's not all. He discovered the elements boron, sodium, aluminum (whose name he later changed to aluminium) and potassium.

In 1860, English physicist **Joseph Wilson Swan** developed the modern incandescent lamp, got his patent and started

THE BOYHOOD OF SIR HUMPHRY DAVY

YOUNG HUMPHRY DAVY MAKING HIS FIRST EXPERIMENTS

Humphrey Davy making his first experiment when a boy. (Courtesy of the Wellcome Library)

manufacturing and selling his bulbs. Swan's house was the first in the world to be lit by electricity. The first bulbs lasted little more than twelve hours but, unlike gas lamps, there was no flame or dirty smoke and they soon grew in popularity.

American inventor Thomas Edison patented his bulb a year later having used much of Swan's ideas and technology. Swan sued in the American courts and won, but he allowed Edison to make and market the bulb in America while the British market was left to Swan, who continued to improve the device. Sadly, it is Edison who is now remembered as the inventor of the electric light, while Swan's endeavours are almost forgotten.

Marine Chronometer

The chronometer is an instrument that accurately measures time and **John Harrison**, a Lincolnshire clockmaker, changed the face of sea travel with his invention. His marine chronometer enabled ship navigators to find longitude when land was no longer in sight.

Accurate navigation at sea has always been essential, but until the invention of the marine chronometer it was not possible. Until Harrison developed his device, ships faced the prospect of getting lost at sea, as captains had to depend on the position of moon and stars to estimate their position. Although this was a simple way to work out longitude, it was sheer guesswork trying to predict the correct local time at any given destination.

For those who don't know what longitude is, it is an imaginary angular line that describes the location of anywhere east or west of a north-to-south line known as the prime meridian. The time difference works out to be one hour for every 15 degrees longitude. This means that each degree of longitude is equal to four minutes.

Understanding these problems faced by seafarers, the government announced a £20,000 prize – worth almost £3 million today – for anyone who could solve the problem.

Harrison decided an accurate clock would do the trick. He set to work to build a portable clock which would not be affected by a ship's movement and that kept the time of any given place on a long sea journey.

Producing a clock like this was difficult because conditions during sea travel vary. The temperature, pressure and humidity changes can cause problems for any timekeeping instrument. However, his marine timekeeper was first tried out on a voyage to Jamaica in 1761 and found to be only five seconds slow. When Captain James Cook set sail on his second voyage of discovery, lasting three years, it was with a Harrison-designed watch. He reported it was never more than eight seconds out.

The Board of Longitude, however, implied that the watch was a fluke and would not be satisfied unless others of the same kind could be made and tested. The government tried to fob Harrison off with a much-reduced sum. Harrison was incensed and made an appeal to Britain's highest authority, King George III, via a letter to his private astronomer at Richmond, Dr Stephen Demainbray. William was duly summoned for interview with the king himself, at which the king is said to have remarked, 'By God, Harrison, I will see you righted!'

Harrison was eventually awarded £8,750 by an Act of Parliament in June 1773. More importantly, perhaps, he was finally recognised as having solved the longitude problem.

It was Harrison's marine chronometer that helped the country achieve pre-eminence as a trading nation.

Radar

Scottish physicist **Sir Robert Alexander Watson-Watt** from Brechin in Angus, Scotland, developed radar to help track storms in order to keep aircraft safe. The means of detecting aircraft by this method provided a vital service in the Second World War as it meant that the relatively small RAF could be directed to the most strategic places. His invention eventually helped the Allies win the war.

The Germans were attempting similar work, but Watson-Watt, who was the Superintendent of the Radio

Research Laboratory in Britain, filed for a patent in September 1935.

The word 'Radar' stands for RAdio Detection And Ranging. Radar is used to locate distant objects by sending out radio waves and analysing the echoes that return. Radar can determine where a distant object is, its size, the speed at which it is moving and its direction. Radar is now used to watch developing weather patterns, monitor air traffic, track ships at sea and for the detection of missiles.

Watson-Watt claimed the invention of radar, but, as with other classic science-based inventions, it evolved. There were precursors and simultaneity of discovery in several countries. Suffice to say that no other government saw the possibilities so clearly and none understood the implications more quickly than the British.

> During the Second World War, British scientists John Randall and Harry Boot invented the magnetron, a device which emits microwaves for use in radar equipment and the source of heat in microwave ovens. Surely a blessing for busy households worldwide.

Fax Machine

Hurrah for Scottish inventor **Alexander Bain,** a farmhand's son from Caithness. Bain, a maker of clocks and instruments, came up with the world's first fax machine in 1846 – the word is short for facsimile – when he used clock mechanisms to transfer images from one sheet of electrically conductive paper to another.

Although Bain received a British patent in 1843 for 'improvements in producing and regulating electric currents

Alexander Bain,
1876, from
*Popular Science
Monthly* volume 9.

and improvements in timepieces and in electric printing and
signal telegraphs' he didn't get much credit for his idea. Why?
Because Samuel Morse claimed it infringed his own patented
telegraph. Several years earlier, Samuel had invented the first
telegraph machine and the fax machine closely evolved from
the technology of the telegraph. The earlier telegraph machine
sent Morse code, dots and dashes, over telegraph wires which
were then decoded into a text message at a remote location.

Frederick Bakewell from Wakefield in Yorkshire is credited
with improving on Bain's invention. He replaced Bain's
pendulums with rotating cylinders that were synchronised,
allowing for a clearer image through better synchronisation.

This primitive device led to what was to become an icon
of the 1980s. For decades before digital technology became
widespread, the scanned data was transmitted as analogue.

Before his fax machine, Bain invented and patented the electric clock in 1840. Then, in 1841, along with John Barwise, a chronometer maker, he took out another important patent describing a clock in which an electromagnetic pendulum and electric current is employed to keep the clock going instead of springs or weights. Later patents expanded on his original ideas.

Photography

The French say that they invented photography thanks to a certain Louis Daguerre in 1834, though the first fixed image was made by Joseph Niépce in 1826 and took eight hours to expose. However, **Thomas Wedgwood,** born in Staffordshire, son of Josiah Wedgwood the potter, beat them to it when he created pictures of insect wings using silver nitrate on leather in 1802.

Daguerre was also in competition with Dorset-born **William Henry Fox Talbot,** the man who invented the Calotype, the development process that became the basis for modern photography and the means of duplicating an image.

In 1835 Fox Talbot made another breakthrough by using silver iodide on paper and discovered how to produce a translucent negative that could be used to make any number of positives by contact printing, a system which was used until the advent of digital cameras. The first ever photograph was thought to be that of a window at Lacock Abbey in Wiltshire produced by Fox Talbot in 1835.

Then there was colour photography. Scot **James Maxwell** of Edinburgh was a physicist who contributed to the field of optics and the study of colour vision. He created the foundation for practical colour photography and presented the world's first colour photograph, a tartan ribbon, in 1861.

Lacock Abbey. (© VisitEngland/VisitWiltshire/Chris Lock/Lacock Abbey)

Maxwell's achievements concerning electromagnetism have been called the 'second great unification in physics' after the first one realised by Isaac Newton.

Text Messages

The world's first text message or SMS (short message service) was sent on 3 December 1992.

Twenty-two-year-old engineer **Neil Papworth** used his computer to wish Richard Jarvis 'Merry Christmas' on his Orbitel 901 mobile phone. Papworth didn't get a reply because there was no way to send a text from a phone. That had to wait for Nokia's first mobile phone in 1993.

The first text messages were free and could only be sent between people on the same network, but in 1994, Vodafone, which was one of only two UK mobile networks, launched a share price alert system. In 1995 the T9 system, which created 'predictive' texting based on letters typed, meant texting could really take off.

Before long, the realisation dawned that there was money to be made from texting. By February 2001 the UK was sending one billion texts a month, which at the standard 10p-a-text charge meant the business was making around £100 million a month.

Text language emerged quickly because of the 160-character constraint of the keypad and also because in the beginning it took some time to type words on a numerical keypad. 'Message' was reduced to 'msg', 'great' was 'gr8' and before 'b4'.

By 2003, exam markers were concerned about text language being used in exam papers. A thirteen-year-old girl wrote an essay in text shorthand, part of which read, 'My smmr hols wr CWOT. B4, we used 2go2 NY 2C my bro, his GF & thr 3 :- kids FTF. ILNY, it's a gr8 plc.' (Or in longhand: 'My summer holidays were a complete waste of time. Before, we used to go to New York to see my brother, his girlfriend and their three (!) kids face to face. I love New York, it's a great place.') Perhaps the new language that texting created is not such a good invention.

Text messaging can save lives:

In 2001, young traveller Rebecca Fyfe was on a backpacking trip in Asia when the boat she was onboard in the Lombok Strait started sinking. She sent an SMS to her boyfriend, 'Call Falmouth Coastguard, we need help, SOS.' The message was passed to coastguards in Australia and then to the Indonesian authorities. A rescue boat took the passengers to safety and the stricken vessel was towed into harbour.

In 2008, Martin Stone got lost after climbing Puig Tomir in Majorca. He sent a text message to his wife in the Midlands, who contacted emergency services. They called the Spanish authorities who launched a rescue operation. Thick fog hampered the search, but Mr Stone was eventually spotted the following morning, waving his red scarf at a rescue helicopter. 'It is only thanks to his wife and his scarf that he's alive today,' said a rescue worker. Personally, I think it is probably thanks to that text message.

Also in 2008, surgeon David Nott was volunteering at a hospital in Congo when a sixteen-year-old boy came in needing emergency surgery. To save the boy's life, the doctor texted a colleague to ask details of the necessary procedure regarding the life-saving amputation to the boy's arm. The colleague, Professor Meirion Thomas, replied with specific instructions which Nott followed successfully. Despite risks involved and lack of intensive care amenities, the boy made a full recovery.

Printing Press

William Caxton, who was born in Kent around 1422, was a businessman, royal advisor, translator and printer who set up England's first printing press in 1476. Caxton had learned about printing in Cologne. He went to Bruges around 1474 and that was where he printed a translation of a French romance, which he entitled *The Recuyell of the Historyes of Troye* and which he finished in 1471. *The Recuyell* was the first book printed in the English language. His second publication, printed in 1476, was *The Game and Play of Chess Moralised*. This was the first printed book on chess and the first to use woodcut illustrations.

Caxton then returned to England and set up England's first printing press at Westminster. There he produced the first book printed in England, called *The Dictes and Sayings of the Philosophers*, in 1477, followed by *Troilus and Creseide*, *Morte d'Arthur*, *The History of Reynart the Foxe* and Chaucer's *The Canterbury Tales*. William Caxton died, almost with pen in hand, around 1491.

Since Caxton would not print regional variations in English, he began the standardisation of the English language and its spelling. This facilitated the expansion of English vocabulary and a widening gap between the spoken and the written word. However, Richard Pynson, who started printing in London in 1491 or 1492 and who favoured Chancery Standard, was a more accomplished stylist and consequently pushed the English language further toward standardisation.

Smallest Mass-Market Transistor Radio/Pocket Calculator

Clive Sinclair was born in July 1940. After a spell in magazine publishing he went on to make a number of ground-breaking inventions.

In 1958, while still at school, he developed the world's smallest mass-market transistor radio. Smaller than a matchbox, it went into production in 1963.

In 1966 he launched the world's first pocket TV, featuring a two-inch cathode ray tube. His next innovation in 1972 was the first pocket calculator. While there had been other attempts to create such a device, the concept depended somewhat on the size of your pocket. However, Sinclair's device was so small, around two inches by five and a half inches and just a third of an inch thick, that it even fitted in the pocket of a shirt. It used an LED screen and tiny hearing-aid batteries.

The LED display was also used in the first mass-produced digital display watch.

Sinclair was also closely involved with the development of home computers via the ZX 80, 81 and Spectrum range, although there were various competitors both in the UK and the US.

In 1984 he produced the first flat-screen pocket TV. Not a flat screen as we know it today – this used a cathode ray tube with an electron gun at right-angles to the screen.

Perhaps many years ahead of its time, in 1985 he produced the Sinclair C5 electric vehicle, a tricycle with a top speed of 15 mph, the fastest anyone was allowed to go without a driving licence.

Clive Sinclair was knighted in 1983.

The ATM

Back in the 1960s, getting your cash out of a bank was not exactly convenient. To withdraw cash, customers had to write a cheque for cash and present it at their bank.

In those pre-computerisation days, bank tellers had to cash up and reconcile branch accounts each day before they went home. This meant that banks closed around 3.30 p.m.–4.00 p.m. in order to give them time to do this. Some were open on Saturday mornings, but trades unions were pressing for them to be closed so that staff could go shopping. Sunday opening of shops hardly existed in those days.

As a result, other retailers cashed cheques for favoured customers outside non-banking hours and those catering for such hours tended to be petrol stations and pubs. However, since there were no plastic cards that could be used for verification, you would have to be known personally to the retailer in order for him to accept your cheque. Of course, it was not deemed polite to pop into a pub just to cash a cheque; customers felt obliged to buy at least one drink while they were there. Thus obtaining cash became a lengthy but sociable event.

The ATM, initially a cash dispenser, was developed to address this problem. Two people, both from Scotland, appear to share credit for the device as we know it today, although complete details are hard to come by, and both were subsequently awarded the OBE.

John Shepherd-Brown was an employee of De La Rue when he came up with the idea of a machine that would dispense cash, thus allowing customers to get money from a bank when it was closed. Apparently he drew his idea from chocolate vending machines. There is little doubt that the first such device was his, following an order placed by

Barclays Bank. They unveiled their first device at a branch in Edmonton, north London, in June 1967.

According to PR Neswire's 'Things You Didn't Know about Barclays Cash Dispensers', the original Barclaycash vouchers were supplied in packs of ten and issued by bank tellers and were free to approved customers only. There was no limit on the number of vouchers a customer was allowed, providing there were sufficient funds in the account.

The vouchers were placed in the left-hand drawer of the 'robot-cashier', as it was known. The drawer was closed and when the machine verified the validity of the voucher, a 'pass' green light would appear. The customer then pressed in their own four-digit number on the keys and a few seconds later another light lit up and £10 in cash was available from the right-hand drawer.

The new device was the subject of an advertising campaign featuring Reg Varney, made famous by his role on the television programme *On The Buses*.

The vouchers were coated with a mildly radioactive substance called Carbon 14, which enabled them to be recognised by the machine. In addition, each user was given a Personal Identification Number (PIN). Originally planned to be six digits long, Shepherd-Brown discovered that his wife could not remember more than four digits, thus the worldwide standard for PINs is four digits. Nevertheless, the process used by Shepherd-Brown's device was never patented for fear it would reveal to fraudsters how the security worked.

In the meantime, the idea of a PIN being stored on a card or voucher by means of a series of mathematical codes, this being the way ATMs work today, belongs to a fellow Scotsman, **James Goodfellow**, and he filed a patent for this in May 1966, a year before Shepherd-Brown's device was launched by Barclays. Subsequently, both NCR and IBM were among those to license this technology.

There is no doubt that later Barclays cash dispensers used eighty-column punched cards, an early means of storing computer data, rather than the Carbon 14-coated vouchers. By contrast, users were initially supplied by post with seven cards in a folder, again each with a value of £10. After voucher number five there was a renewal slip, similar to those used in chequebooks. On receipt of this Barclays posted the user a further five vouchers to replace those that had been used. These devices worked slightly differently. Customers entered their PIN and a drawer slid out of the machine. The card was positioned on it by means of two lugs and the drawer closed. After some whirring and clattering – and checking that the PIN matched the data on the card – the drawer opened to reveal ten £1 notes.

Thus the need to visit the pub on a Friday evening in order to get cash for the weekend became unnecessary. Some might say that perhaps the ATM wasn't such a good idea after all!

Israel has the world's lowest installed ATM at Ein Bokek at the Dead Sea, installed independently by a grocery store at 1,381 feet below sea level.

The Chronograph

George Daniels, born in Sunderland in 1926, was one of the world's finest watchmakers and one of the few modern watchmakers who built entire watches himself.

However, accuracy was one of the problems with mechanical watches. The movement generated friction between components and required lubrication and so required regular servicing. Over time the lubricant thickened, affecting the accuracy of the watch.

Daniels worked on solving this issue, and in 1976 he unveiled the coaxial escapement, an escapement designed in such a way that there was very little friction, therefore the mechanism did not require optimal lubrication. So revolutionary was this invention that it ranks as one of the most important ever in the history of clockmaking and watchmaking.

Thanks to this design, coaxial escapement watches are highly accurate, often more so than quartz watches, and have been known to keep time to within a second or two per month.

His mechanism was adopted by Omega, and in 1999 they launched their first watch using this remarkable development.

And to complete this third section, here are some more great british inventions within the Technology category...

Reflecting Telescope

Invented in 1668 by **Isaac Newton**. When he was a fellow at Trinity College, Cambridge, Newton took the idea of a reflecting telescope and turned it into reality. This huge leap forward in telescope technology made astronomical observation much more accurate.

Typewriter

In 1714, Englishman **Henry Mill**, a waterworks engineer, obtained a patent for a machine that appears to have been similar to a typewriter. The patent says, '[He] hath by his great study and paines & expence invented and brought to perfection an artificial machine or method for impressing or transcribing of letters, one after another, as in writing,

Bletchley Park woman at typewriter, reading document. (©
VisitEngland/Bletchley Park/Shaun Armstrong/Mubsta.com)

whereby all writing whatsoever may be engrossed in paper
or parchment so neat and exact as not to be distinguished
from print; that the said machine or method may be of
great use in settlements and public records, the impression
being deeper and more lasting than any other writing,
and not to be erased or counterfeited without manifest
discovery.'

Steam Hammer

This power-driven hammer for shaping pieces of wrought
iron was invented in 1837 by Scotsman **James Nasmyth.** His
hammers were said to be able to crack the top of the shell
of an egg placed in a wine glass without breaking the glass.
Nasmyth retired at forty-eight saying, 'I have now enough of
this world's goods: let younger men have their chance.' He

pursued various hobbies, including astronomy, and built his own twenty-inch reflecting telescope, in the process inventing the Nasmyth focus, and made detailed observations of the Moon.

Microchip

Geoffrey William Arnold Dummer from Hull was a British electronics engineer and consultant credited with being the first person to conceptualise and build a prototype of the integrated circuit, commonly called the microchip, in the late 1940s and early 1950s.

Radio

Yes, it is true that in 1896 Guglielmo Marconi sent a wireless telegraph over 94 miles, but **David Edward Hughes** from Denbighshire in Wales is recorded as the first person in the world to transmit and receive radio waves. Evans designed the synchronous type-printing telegraph in 1856.

Wind-Up Radio

Invented in 1991 by **Trevor Baylis** after he saw a television programme about AIDS in Africa. The programme said that one way to stop the disease from spreading was for people to listen to educational information on the radio. So Baylis designed one that needed no batteries, running off an internal generator powered by a mainspring wound by a handcrank. He was able to demonstrate it to Nelson Mandela and after that it was distributed throughout the continent.

The iPod

In 2007, computer giant Apple acknowledged the work of a British man, **Kane Kramer**, who came up with the technology that drives the digital media player. His sketches at the time showed a credit card-sized player with rectangular screen and a central menu button to scroll through a selection of music tracks, similar to the iPod. His invention, called the IXI, stored only three and a half minutes of music on to a chip but Kramer rightly believed its capacity would improve. He took out a worldwide patent and set up a company to develop the idea. However, it all came crumbling around him when, after a boardroom split in 1988, he could not raise the £60,000 needed to renew patents across 120 countries. Unfortunately for Kane Kramer, the technology became public property.

4

Medicine

Chloroform

If you ever visit a medical museum you will shiver over fiendish Victorian medical devices. Trepanning tools, gallstone crushers, bone saws, dental keys – in the 1800s surgeons practised in a seriously barbaric way. Most medical practitioners of the day accepted the situation; after all, there was really no alternative. Patients must have dreaded surgical operations. Anaesthetics had not yet been discovered, so patients had to endure excruciating pain. In an amputation, the patient would be held down awake while the surgeon cut through the soft tissue and bone. Surgeons worked quickly, often leading to mistakes and a low survival rate.

The horrors of the operating theatre before the advent of anaesthetics had always haunted obstetrician **Sir James Young Simpson** and preyed upon his mind. He was a kindly soul, gentle and caring, who wanted to cut down the suffering of his patients in childbirth.

James was born in Bathgate, West Lothian, to a working-class family. His first schoolmaster, Mr Henderson, had a wooden leg, and went by the name of 'Timmerleg' among his pupils. The town's weaving population were quite intelligent and 'Timmerleg' was head and shoulders above the usual run of village teachers. He was also a keen naturalist and James Simpson, his brightest pupil, learnt to love nature no less than

Illustration showing use of inhaler file. (Courtesy of the Wellcome Library)

other subjects from the teacher. His intelligence was such that he was known as the wise wean, or the 'wise child'.

He trained to be a doctor and in 1830 passed his final examination with honours. In 1831, when only twenty, he was made a member of the Royal College of Surgeons at Edinburgh. He was too young to take his degree as doctor of medicine, so worked for a time as an assistant in the medical school at Edinburgh.

In 1835 he set himself up in an Edinburgh practice and gradually acquired patients. He is reported to have said, 'They are mostly poor, but still they are patients.' After a year he obtained a hospital appointment, which helped to spread word of his deftness and sympathy.

As time passed it became clear that Simpson could heal where other doctors gave up trying. His charm and brilliance inspired hope and soon he had more patients than he could cope with, coming from all over Europe.

His old Bathgate friends, however, always had first call on his time. When he was not so busy, the password 'An old friend from Bathgate' would open his consulting room door. One day when he was engaged with such a patient, a well-known authoress rang his bell wishing to see Dr Simpson. His servant told her that the doctor could not see any more patients that day. 'But,' persevered the lady, 'I am sure I can be admitted. Take my name, he knows me.' The reply was, 'No doubt, but Dr. Simpson also knows the queen, ma'am.'

He did indeed, and in 1847 he was appointed one of Queen Victoria's physicians for Scotland.

In 1846 the news came from America of the first trial of ether in surgery. No one hailed the discovery more heartily than Simpson, who at once adopted its use. But soon, with his usual prodigious energy, he was seeking out a more reliable anaesthetic than ether, the use of which was often dangerous for the patient. He also disliked it because of its strong smell, which made his patients cough.

Every evening after finishing work for the day, Simpson and his two assistants, Dr Keith and Dr Duncan, inhaled various drugs in the hope of discovering a really satisfactory anaesthetic. On the evening of 4 November 1847, Simpson and his assistants met as usual to try out possible new anaesthetics. Dr Keith started to inhale half a small tumblerful first. A couple of minutes later, he was out for the count, lying unconscious under the table. Moments later, Simpson and Duncan followed him. Anyone coming into the room would have wondered what on earth had happened, perhaps imagining the three men were suffering from the effects of too much alcohol, rather than being in the process of discovering one of the most ground-breaking moments in the history of medicine.

First to wake up, a groggy Simpson knew that he had found something that could be used as an anaesthetic. 'This is far better and stronger than ether,' he declared. 'This will turn the world upside down.'

Mrs Simpson, her sister and her niece had come into the room and watched the scene before them in horror. As soon as Simpson showed signs of returning consciousness, they plied him with anxious questions as to how he felt. The doctor laughed, for he knew now that pain was conquered, that he had stumbled upon a new path of healing.

But it wasn't easy to convince the world. A fierce and dogged resistance arose against the introduction of chloroform. Simpson was assaulted with professional and theological arguments. Not that it worried him; he relished the chance to put forward his case and statistics and demonstrations silenced objectors. Prejudice from within the Church, however, was more difficult to overcome. Chloroform was unnatural, they said. 'So are railway trains, carriages and the steamboats,' retorted Simpson, who organised pamphlets refuting the theologians on their own scriptural grounds, showing that he knew his Bible better than they did. This battle, which now seems so ridiculous, raged for years till Dr Chalmers, foremost of Scottish divines, declared that the question had not even the remotest connection with theology.

Still, intense criticism was directed at Simpson for using chloroform to relieve pain in childbirth. It was not until Dr John Snow – the man who discovered that cholera was a waterborne disease after he traced an outbreak to a Soho water pump – administered chloroform to Queen Victoria during the birth of Prince Leopold that the use of anaesthetic drugs became respectable in this way.

For many years Simpson strove to bring health and happiness to both rich and poor, without too much consideration for his own health. Eventually, crippled with sciatica, he died on 6 May 1870.

Simpson was one of the greatest doctors that ever lived and to his thousands of patients he was 'the beloved physician, who never tired of doing good'.

He wasn't born with the name 'Young' and no one is exactly sure how it came to be included in his name. The general opinion that because he was always the youngest of his colleagues and was therefore known as 'Young Simpson'.

Aspirin

Some discoveries are the brainchild of an individual working alone, while others are the result of teamwork. This could be described as one of the latter.

Vicar **Edward Stone**, born in 1702 in Princes Risborough, Buckinghamshire, was the person who discovered salicylic acid, the active ingredient in aspirin.

One day in 1757, while out for a walk in the country, Edward idly pulled a piece of bark off a willow tree and started to nibble it. His first impression was that it had a bitter taste. Knowing that the bark of the Peruvian cinchona tree from which quinine, used in the treatment of malarial fevers, is derived has a similarly bitter taste, he surmised that the willow might also have therapeutic properties. Stone's interest in willows was due to the ancient 'Doctrine of Signatures' whereby the cause of a disease offers a clue to its treatment.

According to Stone, 'As this tree delights in a moist or wet soil, where agues chiefly abound, the general maxim that many natural maladies carry their cures along with them or that their remedies lie not far from their causes was so very apposite to this particular case that I could not help applying it and that this might be the intention of Providence here, I must own, had some little weight with me.'

He experimented by gathering and drying a pound of willow bark and creating a powder which he gave to about fifty people in the parish of Chipping Norton. It was consistently found to be a 'powerful astringent and efficacious in curing agues'. In fact, Stone had discovered the world's best-selling painkiller and in 1763 he wrote to the Royal Society with his findings.

Not that it was the first time that salicylic acid and its derivatives had been used. The pain-relieving effects of Salix (willow) and Spiraea (meadowsweet) species was known in many cultures. It goes back at least to 400 BC, when Hippocrates prescribed the salicin-rich bark and leaves of the willow tree to reduce pain and fever. In AD 100, Dioscorides mentioned willow leaves as a cure and a hundred years later Pliny the Elder and Galen also mentioned them in medicinal tomes. Although forgotten by doctors in the Middle Ages, it lived on in folk medicine.

Aspirin in tablet form came much later via a German scientist, Felix Hoffman. In 1996 the *New York Times* called it 'the wonder drug nobody understands', and another Englishman, **John Vane**, found that aspirin works by blocking production of a hormone-like substance, prostaglandin, which triggers pain, fever and inflammation. He discovered that even a very low dose of aspirin is useful for stopping formation of blood clots and may help prevent strokes and heart attacks. Vane was rewarded for his findings with a knighthood, numerous fellowships and a Nobel Prize.

Edward Stone was also a JP for Oxfordshire, actively enforcing the Poor Law.

Ibuprofen

In 1939, Stewart Adams from Nottingham left school. He was sixteen and started work at a branch of Boots as a pharmacy apprentice. He learned to mix powders and prepare medicines, and before too long the company sponsored him to attend university in Nottingham where he gained a degree in pharmacy.

He then returned to Boots in 1945 to work on its penicillin production project. Part of the job involved producing penicillin by growing the mould in quart milk bottles. Adams recalled later that there were almost a million bottles on the site. Not long afterwards, he was offered a job in the research department and he gained a PhD at Leeds University in pharmacology. Boots then put him on to a rheumatoid arthritis project. Corticosteroids were used to treat the problem, but unpleasant side effects meant they were not suitable for everyone. So Adams set his mind to look for an alternative non-steroidal compound. Little was known about the disease, and the only non-steroidal drug helpful at that time was aspirin, which was known to possibly cause damage to the gastric lining.

Adams had little progress for a year or two until he read a paper in a dental journal which suggested that aspirin had an anti-inflammatory effect. That was when he began to look for something with a similar power.

In 1956, Adams proposed to Boots that he should be assigned a chemist and that they should begin work on creating a drug that was anti-inflammatory to combat the symptoms of RA: pain and inflammation in the joints, low-grade fevers and flu-like symptoms. The chemist who came to work with him was John Nicholson and he became Adams partner in the scientific discoveries to follow and would be his fellow patentee.

Over the next few years, the scientists made four different compounds that went to clinical trial. The first one failed because it had no effect while the second one was active, as were the third and the fourth, but all had unacceptable side effects including a range of problems from skin rashes to jaundice.

So it was back to the lab and the eureka moment. Adams and Nicholson's fifth compound was ibuprofen, for which they had high hopes. It went through trials and was well tolerated by the body, although very high doses can cause stomach irritation. A number of GPs involved in the trials independently decided to increase the dose and found it was beneficial to their patients. A new non-steroidal anti-inflammatory drug (NSAID) had been born. The original patent was granted for a number of compounds, including ibuprofen, but it took a further seven years for the drug to become licensed for use and available on prescription.

In all it took sixteen years for ibuprofen to progress from an idea to being available, at first on prescription, and Adams was involved every step of the way. The original patent for ibuprofen was granted on 12 January 1962. In 1987 Dr Adams was appointed an OBE for his work in developing ibuprofen.

A range of NSAIDs is now available but recent reports suggest they may increase the risk of heart disease and some of them have dented public confidence. Ibuprofen has emerged relatively unscathed to date; it has fewer toxic properties and is metabolised quickly.

Penicillin

Next time your doctor gives you penicillin, you have **Alexander Fleming,** a modest man from Lochfield in Ayrshire, to thank for helping your recovery.

Fleming studied Medicine at St Mary's Hospital in London, specialising in bacteriology. He researched non-toxic antibacterial and antiseptic substances and in 1928 made the discovery that would change medicine forever.

His lab tended to be rather untidy, and sometimes he forgot to close the windows. Once, before going away for a week, he didn't sterilise all his plates and he left the lab windows open. When he came back to work he noticed that lots of his culture plates had mould growing on them. He noted that surrounding the mould was a clear halo, mould-free. Left for a further period, more mould grew in the dish, but never over this clear area. This meant no bacteria were growing here, so something must have killed them. At the time, he described the discovery as a 'triumph of accident and shrewd observation'.

Unlike scientists John Tyndall and D. A. Gratia, who had noted the phenomenon years earlier, Fleming recognised just how important his discovery was. He understood a substance which prevented bacterial growth was produced by the mould. He named it penicillin.

His findings were published in the *British Journal of Experimental Pathology* in 1929 but for ten years not much interest was shown in his discovery. Howard Florey, an Australian professor of pathology, and Ernst Chain, a German-born professor of biochemistry, led the team reinvestigating penicillin. Work started on small-scale manufacture of the drug, which enabled it to become widely available. It was found to be effective against many serious ills of the day such as tuberculosis, scarlet fever, pneumonia, diphtheria, gangrene and syphilis in addition to wound and childbirth-related infections. Since then it has saved millions of lives.

However, as Fleming himself noted, if insufficient doses were administered, resistant strains of bacteria developed, which is still a problem today.

At the onset of the Second World War, enormous efforts were made to ensure the mass production of penicillin by US pharmaceutical companies and by the end of the war enough had been produced to treat all soldiers who needed it.

Both Fleming and Florey received knighthoods in 1944 and the Nobel Prize for Physiology or Medicine in 1945, along with Chain. Fleming was given numerous other awards including the Medal for Merit (USA 1947), the John Scott Medal from the City Guild of Philadelphia (1944), the Honorary Gold Medal of the Royal College of Surgeons (1946) and the Grand Cross of Alphonse X the Wise (Spain, 1947). He was even named Honorary Chief Doy-gei-tau of the Kiowa tribe.

> Although the name of Sir Alexander Fleming has become synonymous with the discovery of penicillin, the other key players sadly remain virtually unknown.

Hypnotism

James Braid, born in Kinross, Scotland, in 1795, was a surgeon, a scientist, a significant innovator in the treatment of club-foot and an important and influential pioneer of hypnotism and hypnotherapy. He is regarded by many as the first genuine 'hypnotherapist' and the 'father of modern hypnotism'.

Braid's interest in the subject began in 1841 when the exotic Swiss mesmerist Charles Lafontaine visited Manchester to give performances. He always dressed in black and had a very long beard. People didn't really know what to make of him and thought he was either crazy or a charlatan, but they flocked to see his shows anyway.

Braid was one of those who suspected Lafontaine of being a fraud and his initial expectations seemed to be confirmed. However, six nights later at another exhibition, Braid, impressed by one patient's inability to open his eyelids after being hypnotised by Lafontaine, wrote, 'I considered that to be a real phenomenon and was anxious to discover the physiological cause of it. Next night, I watched this case when again with intense interest, and before the termination of the experiment, felt assured I had discovered its cause, but considered it prudent not to announce my opinion publicly, until I had had an opportunity of testing its accuracy, by experiments and observation in private.'

Being the kind of person he was, he was determined to try and find a scientific reason for the trance. He started experimenting. Believing that the 'sleep' resulted from fatigue of the eyes, Braid used his wife, a friend and a servant as guinea pigs. When all of them were instructed to gaze steadily at an object, he discovered he too could produce a trance-like state.

At first his technique was to hold a small, bright object between eight to sixteen inches in front of his subject's eyes so that the eyes became strained, after which the eyelids would often close spontaneously. As he continued with his experiments, however, he found he achieved trance states by suggestion alone.

In 1842 he published *Neurypnology or The Rationale of Nervous Sleep Considered In Relation With Animal Magnetism*. Having concluded that the phenomena was a form of sleep, Dr Braid named the phenomena after Hypnos, the Greek god of sleep and master of dreams.

Hip Replacement

Surgeon **Sir John Charnley** from Lancashire made a great step forward in medicine when he offered new hips for the old to thousands of people. To be fair and accurate, he was not the first person to have the idea; the first hip replacements took place in Germany in the 1890s with ivory and gold used in place of the ball on the femur, though they were neither successful nor long-lasting. In September 1940 an American surgeon, Dr Austin Moore, performed the first metallic hip replacement using an alloy of cobalt and chrome known as vitallium. A later version was introduced in 1952 and is still in use today, although only rarely.

In 1960 a Burmese surgeon, Dr San Baw, at the Mandalay General Hospital, used an ivory prosthesis to replace fractures in the neck of the femur and between 1960 and 1980 is reported to have used over three hundred ivory hip replacements. It is believed ivory was popular in part because in Burma it was cheaper and more accessible than metal.

However, it was **Charnley**, an orthopaedic surgeon, who pioneered the modern hip replacement surgery that has become a global standard.

Back in 1935, Charnley was a house surgeon at Manchester Royal Infirmary. He became a fellow of the Royal College of Surgeons the following year. In 1937 he moved to Salford Royal Hospital, returning to Manchester in 1939. Following the outbreak of the Second World War he volunteered to join the Royal Army Medical Corps and was subsequently posted to Dover, where he was involved in the evacuation of Dunkirk. After that he was transferred to Cairo where he spent the majority of his service career.

His work as a surgeon flourished but he wanted to pursue his interest in the lubrication of joints. After the war, and having returned to Manchester, he handed over most of his clinical work in order to get time to set up a hip surgery centre at Wrightlington Hospital in Lancashire. There he continued his research into joint lubrication. Many surgeons at the time believed that a film of fluid was responsible for the low friction between two joints; Charnley disagreed, and his experiments proved he was correct.

In the early 1950s, Charnley had examined a patient who had been fitted with a replacement hip made from acrylic plastic. It was very noisy. In fact, the acrylic hip squeaked so loudly that the man's wife tried to avoid being in the same room as him.

This led Charnley to look for a slippery substance to use in the socket of a hip replacement. He initially chose Teflon, which he began to use around 1960. However, within a year or so it became clear that Teflon was not suitable as it began to show signs of wear and had other drawbacks.

Determined to find a better material for use in his operations, Charnley came across a substance called high-molecular-weight polyethelene, which he first used in 1962. He spent several years observing the results of his surgery and the performance of the material but no problems were found.

To this day Charnley's procedure is the standard by which all other procedures are judged, and thanks to his work hip replacements are not just for those younger patients with hip injuries but also for older patients and recovery to full mobility is fairly rapid.

Hip replacement surgery is now one of the most common operations to be carried out worldwide and is regularly used to relieve the pain of arthritis or damage to the hip itself.

Smallpox Vaccine

It is a disease we need fear no longer, thanks to **Edward Jenner**, a physician from Gloucestershire. The work of the pioneer of the smallpox vaccine, the world's first, is said to have 'saved more lives than the work of any other human'.

At the age of fourteen Edward was apprenticed for seven years to surgeon Mr Daniel Ludlow. In 1770, he became apprenticed in surgery and anatomy under surgeon John Hunter and others at St George's Hospital in London. Hunter gave Jenner William Harvey's advice, famous in medical

Jenner and his two colleagues seeing off three anti-vaccination opponents. (Courtesy of the Wellcome Library)

circles and characteristic of the Age of Enlightenment, which was, 'Don't think; try.' Before long, Jenner became a successful doctor and surgeon.

Inoculation was already a standard practice but in those early days involved serious risks. In 1765, Dr John Fewster published a paper in the London Medical Society entitled 'Cow Pox and Its Ability to Prevent Smallpox', but he did not pursue the subject further.

Over the following years investigators successfully tested a cowpox vaccine in humans to defeat smallpox. Dorset farmer Benjamin Jesty vaccinated his wife and children during a smallpox epidemic in 1774, but it was not until Jenner's work twenty years later that the procedure became understood. Indeed, Jenner may have been aware of Jesty's procedures and success. The initial source of infection was a disease of horses called 'the grease', which was transferred to cattle by farm workers and manifested as cowpox.

Jenner knew that milkmaids were generally immune to smallpox, so he guessed that the pus in blisters that milkmaids received from cowpox, a disease similar to smallpox but much less virulent, gave them protection from smallpox.

On 14 May 1796, Jenner scraped pus from cowpox blisters on the hands of Sarah Nelmes, a milkmaid. She had caught cowpox from a cow called Blossom. Incidentally, Blossom's hide hangs on the wall of the St George's Medical School library. Jenner tested his hypothesis by inoculating his gardener's son, eight-year-old James Phipps, on both arms. Phipps was the seventeenth case described in Jenner's first paper on vaccination. Poor James took a fever due to the vaccination, but suffered no full-blown infection. Later, he injected the little boy with variolous material, the routine method of immunisation at that time. Again, fortunately for James, no disease followed.

Dr Donald Hopkins wrote, 'Jenner's unique contribution was not that he inoculated a few persons with cowpox, but that he then proved by subsequent challenges that they were immune to smallpox. Moreover, he demonstrated that the protective cowpox pus could be effectively inoculated from person to person, not just directly from cattle.'

Jenner continued his research and after some time published his findings. Some were correct, others were not. The medical establishment deliberated over his report before accepting it. Eventually vaccination was accepted, and in 1840 the British government banned variolation – the use of smallpox to induce immunity – and provided free vaccination using cowpox.

The success of his discovery spread around Europe and was used in the Spanish Balmis Expedition, a three-year-long mission to the Americas, Philippines, Macao, China and Saint Helena to vaccinate thousands of people. The expedition was successful and Jenner wrote, 'I don't imagine the annals of history furnish an example of philanthropy so noble, so extensive as this.'

Jenner's continuing work on vaccination took up so much of his time that it prevented him from continuing his ordinary medical practice. However, after petitioning Parliament he was granted £10,000 for his work. In 1803 he became involved with the Jennerian Institution, a society concerned with promoting vaccination to eradicate smallpox and in 1806 he was given another £20,000 for his continuing work in microbiology.

In 1811 Jenner observed a significant number of cases of smallpox after vaccination. He found that the severity of the illness was notably diminished by previous vaccination. In 1821 he was appointed Physician Extraordinary to King George IV.

In 1979, the World Health Organisation declared smallpox an eradicated disease. This was the result of coordinated public health efforts by many people, but vaccination was an essential component. Jenner's vaccine also laid the foundation for contemporary discoveries in immunology, and the field he began may someday lead to cures for arthritis, AIDS and many other diseases.

Antiseptic Surgery

Joseph Lister was a surgeon who transformed surgical practice in the late 1800s by promoting carbolic acid as an antiseptic. He is regarded as the founder of antiseptic medicine and the inventor of preventive medicine.

Born in Essex, he was interested in surgery from an early stage and attended the first surgical procedure carried out under anaesthetic in 1846. He was concerned that after undergoing surgery several patients died from unattended infections. In those days surgeons believed that infections came from the wounds themselves. They did not bother to wash their hands or change their bloodstained clothes. Lister thought differently. After reading Pasteur's work on microorganisms, he decided to experiment using one of Pasteur's proposed techniques, that of exposing the wound to chemicals. He started cleaning patients' wounds with carbolic acid and used dressings soaked with it to cover the wound and noted that the rate of infection was reduced. He conducted research and spent years trying to discover how infections could be stopped. He washed his hands after every surgery, sterilised instruments, wore clean clothes and sprayed carbolic acid in the operating theatre to limit infection. This approach, which of course is commonplace nowadays, led to fewer deaths from infections among patients.

Listerine mouthwash was named after him. Originally the mouthwash was used as an all-purpose antiseptic.

Steri-Spray

Forty-seven-year-old **Ian Helmore** from Hertfordshire left school at fifteen with no qualifications and started work at a carpet warehouse. For the next seven years he progressed through the ranks to fitting carpets and during this time also helped out at his father-in-law's water treatment company. He enjoyed working there and eventually became a permanent employee. With a desire to learn more, he went to evening classes where he studied for a City & Guilds Plumbing qualification.

The business was sold when Ian's father-in-law retired, but Ian continued to work as the company's chief engineer before leaving in 1994 to start his own business, Helmore Industrial Water Treatment.

Ian Helmore inventor of Steri-Spray. (© Chris Helmore)

While working in a hospital during this time, he became aware of Legionella bacteria in water and how legionnaires' disease could be caught from infected showers. Legionnaires' is a potentially fatal form of pneumonia which is contracted by breathing in droplets of water from contaminated water systems which contain Legionella germs. The presence of such bacteria in showers was discovered at a hospital Ian was working at, meaning the whole hospital had no choice but to function with only two working showers. To Ian this was an unacceptable situation, which prompted him to find a solution.

Studies of various types of water treatment led to the idea of inserting ultraviolet (UV) into the shower head. With a prototype made, Ian set about getting his idea patented. Attempts to use the route provided by invention promotion companies proved fruitless, as did gaining a licence with a major $13 billion US corporation. The negotiations and huge worldwide patent fees involved proved to be insurmountable, and Ian was forced to remortgage his house twice. Ian's wife Gill supported him and co-ran the water treatment business while Ian pursued his venture. Eventually her loyalty and their perseverance would start to reap rewards as they began the journey to *Dragons' Den*.

An argument with his teenage son Chris led Ian to an ultimatum – he had until his fortieth birthday to get his idea off the ground. A sharp dose of teenage realism stating that the shower idea might never come to anything prompted Ian to apply to the BBC's *Dragons' Den* programme. His subsequent appearance in the fifth series in 2007 attracted the interest and investment of Theo Paphitis and Deborah Meaden.

Steri-Spray was finally born.

The product and its components are all manufactured in the UK. Further tests have been carried out on the product by the NHS, and Steri-Spray now counts the NHS and other members of the public sector as valued customers. Tests were carried out specifically by the Belfast NHS Trust. While incoming water supply to showers was found to contain legionella, the water coming from the Steri-Spray showers was declared clear of the bacteria. The system was a proven success.

As a result, Steri-Spray showers are being used in high-risk cancer wards. The product has been crucial to stopping legionella bacteria coming into contact with patients whose immune system would not be able to fight off such a disease.

The inventing didn't end there. With the showers a success, Ian was alerted to the sad cases of five infant deaths which occurred in three separate special care baby units in the UK. This was caused by a different bacteria called *Pseudomonas aeruginosa* and was traced back to the water from taps. Ian set about making a tap that could also incorporate UV. Now these new taps have been installed into special care baby units around the UK. They have proved to be hugely successful. Results from independent testing by the Health Protection Agency showed them to be 99.9999999 per cent effective in killing this bacteria.

Portable Defibrillator

Frank Pantridge, born James Francis Pantridge, has saved the lives of many thousands. He served as a medical officer during the Second World War but was captured when serving in the Far East. He was sent to work in the slave labour camps on the Siam–Burma Railway and believed he lived to fight another day only because of the atomic bombs which hit Japan.

Born in 1916 in Co. Down, he graduated in Medicine from the Queen's University of Belfast in 1939.

Many British prisoners of war died of fatal cardiac beriberi. Back in Britain, Pantridge concentrated on cardiac disease. He knew that most coronary deaths result from ventricular fibrillation, which is a disturbance of heart rhythms. This can be corrected by applying an electric shock of momentary duration on the chest but only if the patient gets help in time, which at that time most did not.

Pantridge created a portable defibrillator, powered it with car batteries and put it in an ambulance. This pre-hospital coronary care unit is known as the Pantridge Plan. On the mobile defibrillator he incorporated a fail-safe mechanism to ensure the defibrillator does not deliver a shock unless the lethal arrhythmia is present.

Pantridge's defibrillator is in worldwide use.

Identifying the Mosquito as the Carrier of Malaria

Patrick Manson studied Medicine at Aberdeen University and after he qualified he went to China to work with the Imperial Customs Service. He was one of the first clinicians to introduce vaccination there. However, it was during his research on tropical diseases that he discovered that the mosquito is the host to a developing parasite that causes filariasis, the invasion of body tissues by a worm. Malaria is a worldwide and sometimes deadly disease causing chills, fever, anaemia and enlargement of the spleen.

Enter **Ronald Ross**, born in India to British parents. He was sent to Britain for schooling when he was eight years old and grew up to become a doctor. After qualifying he returned to India as a medical service officer. It wasn't long before he

----HEALTH HINTS----

A N army marches on its stomach, but it won't march very far unless what goes into its stomach is of the right sort.

If the food—liquid or solid —is contaminated with disease-producing bacteria, diarrhoea or one of the more serious complaints such as typhoid, dysentery or cholera will be an almost certain sequel.

In North Africa dysentery is very prevalent amongst the natives. They infect the food that they handle and so infect you if you are not careful of what goes into your mouth.

Would you eat an apple that was crawling with maggots ? Of course you wouldn't. Therefore don't eat any fruit or vegetables uncooked (unless they are skin fruit like oranges or lemons), for they will be undoubtedly crawling with millions of disease germs, which you can't see.

Ice-cream, raw milk and mineral waters must also be denied you, however great

the temptation on a hot day. They are likely to be loaded with dangerous bacteria.

Finally, never drink water *anywhere* unless it is boiled or chlorinated.

We know it's a bit hard, but try and remember that conditions out here are vastly different from at home and *we have got to keep fit.*

The Army diet is carefully chosen by experts who have worked out the proteins, fats and vitamins and all the other funny things which are necessary for good health. It is quite ample for all your needs, but if you want to buy anything extra get some lemons : eat one every day if you like—you soon get used to 'em.

Remember, a little self-denial will become a big factor in getting you home quicker, and then you can gorge yourself with ice-cream, fish and chips, and stone ginger to your heart's content.

Health hints against malaria. (Courtesy of the Wellcome Library)

became convinced that mosquitoes breed in water, which was of personal importance since he had water butts outside his bungalow.

In 1894 Manson published his mosquito-malaria hypothesis, which suggested the mosquito played host to a malarial parasite, passing the parasite on to human beings through its bite. Ross and Manson met and discussed the issue. After Ross returned to India he discovered the malarial parasite in the gastrointestinal tract of the Anopheles mosquito.

It was the breakthrough they needed to combat malaria. By the 1960s malaria was almost eradicated in former British colonies with the elimination of standing water, the use of DDT to destroy mosquitoes and chloroquine, a synthetic form of quinine.

In the 1890s Manson returned to London, where he established the London School of Tropical Medicine. In 1902, Ross received the Nobel Prize for Medicine for his work.

In Hong Kong Manson established a medical school that became the University of Hong Kong.

And to complete this section, here are some more Great British Inventions within the Medicine category...

Hypodermic Syringe

It is generally accepted that the syringe was invented in 1853 by Scot **Alexander Wood** from Fife. While the syringe itself has been known since ancient times, Wood's innovation was to design a syringe that would allow drugs to be administered intravenously without first having to cut the skin. Apparently he found inspiration from the sting of a honeybee. However, Irishman Francis Rynd is recorded as having given medication to his patients for pain via injection in 1844. Who knows why he didn't publish his results until 1861?

ORIGINAL HYPODERMIC SYRINGE OF
DR. ALEXANDER WOOD

THE FIRST USED IN GREAT BRITAIN

Original hypodermic syringe of Dr Alexander Wood. (Courtesy of the Wellcome Library)

MRI Scanner

Created by a team led by **John Mallard,** Scotland's first professor of medical physics at the University of Aberdeen. The first whole-body MRI scan (the initials stand for 'magnetic resonance imaging') was carried out at Aberdeen in 1980. Huge development costs and safety fears hampered its initial progress and rival versions from the USA began to appear. However, it is safe to say that Mallard and his team deserve principal credit for the MRI machine.

Gray's Anatomy

Published in 1858, this enormous digest is the definitive anatomical reference work. **Henry Gray** and **Henry Vandyke Carter** are responsible for its creation. The surgeons from St George's Hospital, London, 'began collaborating to produce a practical anatomy textbook for their students in 1855'. They performed dissections together for eighteen months and Gray wrote the text with Carter making the precise illustrations. The book was a huge success. Carter sailed for India and a career in tropical medicine but sadly Gray died not long afterwards of smallpox. In 1884, Carter was instrumental in establishing a course for women doctors at Grant Medical College.

Blood Transfusion

James Blundell, who was born in London in 1790, was the first doctor to practise blood transfusions. In 1818 he gave a woman a blood transfusion when she started to haemorrhage after giving birth. He used blood from her husband. In all, he performed ten blood transfusions during his career, five of which benefited the patients.

5

Sport

Golf

Although the Dutch talk of a thirteenth-century sport called 'colf' and the French say they first had the idea with 'palle-mail' in the 1400s, it is the Scots who have been most widely credited with having invented the game of golf. The first written record of golf is James II's banning of the game of 'gowf' in 1457 because he wanted his troops to concentrate on archery.

This may upset the Chinese, of course, who were playing a game similar to golf 1,000 years ago. They had a game called *chuiwan* – '*chui*' meaning to hit and '*wan*' meaning ball. Players used ten clubs, including a *cuanbang*, equivalent to a driver today, and a *shaobang*, a three wood or spoon. A description of the sport, written during the Song Dynasty (AD 960–1279), has been found in a volume called the *Dongxuan Records*. In the book, a Chinese magistrate instructs his daughter 'to dig goals in the ground so that he might drive a ball into them with a purposely crafted stick'.

However, all this aside it is generally accepted that the first place where the modern aspects of the game were brought together was in Scotland. Scots were also the first to use holes rather than targets. Archbishop Hamilton's Charter dated 1552 formerly allowed 'golff, futball, schuteing and all gamis' on the Old Course at St Andrews, recognised by Guinness

Old Course, St Andrews. (© VisitScotland/Scottish Viewpoint/P. Tomkins)

World Records as the world's oldest golf course. Golf is also documented as being played on another very old course at Musselburgh Links, East Lothian, and Mary, Queen of Scots, played at least one game of golf there in 1567.

> The oldest surviving rules of golf were compiled in March 1744 for the Company of Gentlemen Golfers, later renamed the Honourable Company of Edinburgh Golfers. They drafted 'Articles and Laws in Playing at Golf', known as the first scribed rules of golf.

Cricket

It is not entirely clear when cricket was first played, though records indicate that as long ago as 1301 the then Prince Edward played a game called craeg, which some believe to be an early form of the game.

The earliest known reference to the game, however, was in evidence given at a court hearing that 'creckett' was played in Guildford, Surrey, in 1550.

The game was initially believed to have been a children's game but there are indications early in the seventeenth century that adults had taken it up, with references to inter-parish cricket. While not inherently a dangerous game, records show that in 1624 Jasper Vinall became the sport's first recorded fatality when he was hit on the head by a ball.

By the end of the century, the game was well established in the south-east of England and was the subject of significant gambling. By the early 1700s the game was extremely popular and large crowds would turn up to see games at the Artillery Ground in Finsbury. While it was certainly not the first game played there, records indicate that on 31 August 1730 a match took place between London and Surrey, which London won.

At that time, a single-wicket version of the game was very popular. The game was played between two individuals who took it in turns to bat or bowl. The bowler was assisted by a team of fielders. For some time during the mid-eighteenth century, the game was more popular than the eleven-a-side version. It was the subject of much gambling and even accusations of match-fixing.

Games were also regularly played at White Conduit Fields in Islington, and it was considered largely a sport for aristocrats and other elite citizens. So fed up were they about the crowds of people who came to see them play, they asked Thomas Lord, a bowler with White Conduit Cricket Club, to set up a new private ground. He leased land on Dorset Fields, now known as Dorset Square, hence the name Lord's, where he staged his first match in 1787, Middlesex versus Essex. This marked the beginning of the Marylebone Cricket Club (MCC).

A year later, the MCC set out a code of laws and this code was adopted by all other clubs. These included using three stumps as a wicket instead of the earlier two stumps between which a ball could pass, and the leg before wicket rule. To this day the MCC is the custodian of the sport's rules. Lord's moved to Regent's Park between 1811 and 1813 before moving to the site of an old duckpond in St John's Wood in 1814, where it has remained ever since. Perhaps this is the origin of the phrase 'out for a duck', but this is pure surmise on the part of the author!

In those early days the pitch was prepared by allowing sheep to graze on it, but in 1864 the club appointed its first groundsman and acquired its first mowing machine. In 1887, in an effort to get involved in county cricket, Middlesex was invited by Lord's to use it as their home ground.

Development of the game continued with underarm bowling being replaced by overarm, and the game was developed at county level, with the county championship starting in 1890.

The game also spread across the British Empire and it grew in popularity in North America, Australia, South Africa and the Caribbean. The first international match took place in 1844 between Canada and America and in 1859 an England team departed on its first overseas tour, to North America. Three years later an England team made the first tour of Australia and in the Australian summer of 1876/7 England played the first ever Test match against Australia at the Melbourne Cricket Ground.

Strong rivalry between the two nations resulted in a series of Test matches called the Ashes. It is believed that the term originated in a newspaper called *The Sporting Times* which, following the defeat of England by Australia at the Oval in 1882 (their first win on English soil), wrote that cricket had died, the body would be cremated and the ashes be taken to Australia. Thus Ivo Bligh, the captain of the English team playing in Australia in 1882–3, said he would 'regain those ashes'. When England had won two of the three Test matches on that tour, a small urn was presented to Bligh by a group of Melbourne women, one of whom Bligh went on to marry. The urn purported to contain the ashes of an Australian wooden bail and some year later Bligh's widow presented it to Lord's. The urn is kept in the museum at Lord's, regardless of which side wins it, although it has been sent to Australia to mark special occasions.

The game of cricket is the second-most popular game in the world, second only to football.

Started in 1842, Canterbury Cricket Week is the oldest cricket festival in England. It involves a series of consecutive Kent County Cricket Club home matches, usually held in the first week of August.

Cricket match. (© Pawel Libera/Visitlondonimages/Britainonview)

Darts

For centuries people have thrown things at wooden targets, and any of these examples could be considered as the origin of darts. It certainly featured as a fairground attraction in a number of countries, where small spears were thrown at a target with concentric circles.

However, the game as we know it today is all about the layout of the dartboard. Although there were regional variations in the layout, some ascribe the version we know today to a carpenter called Brian Gamlin in 1896, though there is no concrete evidence to support Brian's role in the invention of darts. There is stronger evidence, though, to support that the layout was the creation of craftsman and domino maker Thomas William Buckle in 1913, although of course the game was played long before that, using dartboards with different layouts and numbering schemes.

Darts is by far the most popular game in pubs nowadays, though at one time games of chance were banned in pubs. In 1908 publican Jim Garside of the Adelphi Inn in Leeds was hauled before the courts for allowing games of chance to be played in his pub. The court agreed that a Mr Annakin, the pub darts champion, could set up a dartboard in the court and he threw three darts into the 20 segment. Court officials were invited to do the same but were unable to do so. The court had no option but to agree that darts was a game of skill.

In order to increase sales, brewers started to organise darts leagues around 1925, and in 1927 the then largest circulation Sunday newspaper, the *News of the World*, organised a championship, although it was largely a London affair. Nevertheless it drew large crowds and made the game more popular.

In 1954 the National Darts Association of Great Britain was formed, its main goal being to attract sponsorship and run competitions. In 1973 the British Darts Organisation was formed and in 1978 it organised the Embassy World Professional Darts Championship, one of the biggest events in the sport.

Darts began to attract more television coverage and with it the opportunity to earn more money as a professional player.

By the end of 1992, sixteen of the top darts players, keen to generate more earning opportunities, formed what was to become the Professional Darts Association. They were subsequently banned by the BDO from participating in the Embassy tournament after the 1993 event. Embassy supported the tournament until 2003, when the government's ban on tobacco sponsorship of sports events came into force. In 2005, Sport England officially recognised darts as a sport.

Despite the schisms at professional level, darts is still a hugely popular and sociable sport, now played by millions of people in many countries round the world.

> One hundred points in darts is called a Ton. A Ton 80 (180 points) is three triple-twenties.

Football

As a generic term, the word 'football' applies to a number of games in which a ball is kicked. Such games date back to the ancient Greeks and Romans, but the modern games, each of which has a code and set of rules, are a much later invention.

In the Middle Ages, teams of unlimited numbers, each representing a village or town, would kick a ball made of an inflated animal's bladder. These games often occurred on Shrove Tuesday and the games were often referred to as mob football. The 'mob' or team would follow the ball, and should the player with the ball lose it then it was up to the rest of the team to try and win the lost ball. Hardly the elegant game we know today.

These early versions of the game were not always popular. In 1314 the then Lord Mayor of London issued a decree banning the game, as did Edward III in 1363. A similar game, played in Scotland, was prohibited by the Football Act 1424 which, although it fell into disuse, was, amazingly, not repealed until 1906. Indeed, there were many attempts to ban the game because of the unruly nature of participants and spectators as well as the damage it caused.

Various forms of the game developed in public schools, and scholars and teachers took to setting out codes or rules so that teams from other schools could play each other. One of the earliest references to such games is in a book called *Vulgaria*, written in 1519 by William Herman, who had been a headmaster at Eton and Winchester. The book contains the sentence, 'We wyll playe with a ball full of wynde.'

Schools developed their own versions of the game. One version favoured the ability of players to carry a ball, and the most famous of these schools was Rugby, from where the modern game came. Schools such as Eton, Harrow and Westminster favoured kicking and dribbling of the ball.

However, the rapid growth of football was the result of two developments. One was the widespread growth of two codes – the kicking and the carrying versions – and the other the development of the lawnmower, which made it possible to prepare a proper pitch on which the games could be played.

According to FIFA, the real foundation of the modern game was as recent as 1863, when rugby football and association football went their separate ways. It was not until 1869 that not only carrying but handling the ball became outlawed.

However, according to Patrick Barclay, writing in the *Independent* in August 2013, it was the Scots who transformed the unattractive mob-style 'kick-and-rush' form of football into the attractive and stylish passing game it is today. According to Barclay, young men, mostly from the Scottish Highlands, got together at Queen's Park in Glasgow in 1867. They obtained a copy of the FA laws and amended them to meet the style of passing and dribbling that was more suited to them. This style of play suited the Scots because it was less physical than the English game and the Scots, by and large, were of smaller stature, so they used their style of play to gain advantage over the English.

The Football Association's website informs that English football's governing body was formed in 1863. 'Organised football' or 'football as we know it' dates from that time.

Ebenezer Morley, a London solicitor who formed Barnes FC in 1862, could be called the 'father' of the association. He wasn't a public school man, but old boys from several public schools joined his club and there were feverish disputes about the way the game should be played.

Wembley Stadium, London. (© VisitEngland/Diana Jarvis)

Morley wrote to *Bell's Life*, a popular newspaper, suggesting that football should have a set of rules in the same way that the MCC had them for cricket. His letter led to the first historic meeting at the Freemasons' Tavern in Great Queen Street, near the current site of Holborn tube station.

The FA was formed there on 26 October 1863, a Monday evening. The captains, secretaries and other representatives of a dozen London and suburban clubs playing their own versions of football met 'for the purpose of forming an Association with the object of establishing a definite code of rules for the regulation of the game'.

Once the rules were codified, adoption of the game was rapid. By 1871 the Football Association had fifty member clubs. In 1872 the first competition, the FA Cup, was established and the first international match, England versus Scotland, was played in Glasgow on 30 November 1872, this despite the fact that the game was hardly known internationally and the Scottish Football Association was not formed until three months later.

William McGregor, an Aston Villa committee man and FA stalwart, recognised that football had an urgent need for an organised system of regular fixtures involving the top clubs. He wrote to some of those clubs about a league format for football, and a 'Football League' with twelve clubs came into being after just two meetings in 1888. The FA was still the ultimate authority, but the league would live as a self-contained body within it.

At the time most players were amateurs, but in 1879 a small Lancashire club, Darwin, managed to hold the supposedly unbeatable Eton to two draws in the FA Cup of 1879 before finally losing at the third attempt. Two of Darwin's players, John Love and Fergus Suter, are reported as being the first players ever to receive remuneration for their football talent. Professionalism was legalised by the FA in 1885.

The game spread internationally, much as a result of British international influence. Following Netherlands and Denmark in 1889 were New Zealand, Argentina, Chile, Switzerland, Belgium, Italy, Germany, Uruguay, Hungary and Finland in 1907.

The world governing body, FIFA, was formed in Paris in 1904 with just seven members. Between 1937 and 1938, the modern-day laws of the game were set out by future FIFA president, Stanley Rous, who refereed the FA Cup in 1934. He took the original laws, written in 1886, made some alterations and redrafted them in order.

The first World Cup was held in Uruguay in 1930, but the FA didn't enter a team. England had only played against the other home associations before making a tour of Austria, Hungary and Bohemia in 1908. The FA had withdrawn from FIFA, the competition's organisers, over a disagreement regarding payments to amateur players.

Following the end of the Second World War, and with the spirit of 'Europe' that it generated, the FA rejoined FIFA. It was Stanley Rous's suggestion that a Great Britain team played a celebration match against the Rest of Europe in Glasgow, with the £30,000 proceeds going to help refinance FIFA. The rest, as they say, is history.

So the game of association football or soccer (that being an abbreviation of association and thus differentiating it from rugby football) as we know it today is very much a British product, a combination of English invention and Scottish refinement, and without doubt the biggest game in the world.

Perhaps **Lee Jenkins'** Soccer Trolley could get a mention here. The Aberystwyth Town FC academy coach invented a box for transporting football training equipment that could hold a first aid kit, a notepad, pens, six balls, marker cones, a ball pump, a whistle and training bibs, with a magnetic dry-wipe tactics board on top. It was displayed at an international exhibition of inventions at Geneva in 2010. Mr Jenkins is reported as saying,

'It's lightweight but very robust. I've been using it for the past twelve months and the biggest test was when I unveiled it to the youngsters I was training and it proved to be a big hit.' The former Aston Villa, Port Vale and Birmingham City player received support through the Wales Innovators Network, an EU-funded service from the assembly government to help budding entrepreneurs and inventors.

Peter Shilton has made the most appearances in English League football. The keeper racked up an astonishing 1,005 League appearances.

Rugby Football

As mentioned, the two codes, association football and rugby football, went their own separate ways in 1863, but the origins of rugby football appear to go back to 1823, when a William Webb Ellis, playing football at Rugby School, long before the two different games were codified, picked up the ball and ran with it.

The claim was made in 1876, but it seems there is not much hard evidence to support the fact, especially as Ellis had died in 1872. Nevertheless, Webb Ellis, who later became an Anglican clergyman, survives as the founder of the game. There is a statue of him outside the school and also a plaque which reads, 'This stone commemorates the exploit of William Webb Ellis who with a fine disregard for the rules of football as played in his time first took the ball in his arms and ran with it thus originating the distinctive feature of the rugby game A.D. 1823.'

The William Webb Ellis Cup was also named after him and is presented to the winner of the Rugby World Cup.

Following the split between association football and rugby football in 1863, the first rugby football clubs were founded and in 1871 twenty-one clubs met in London to set up the Rugby Football Union. Over the next few years the rules were modified. The number of players in a team was reduced from twenty to fifteen, and a player held in a tackle was required to put down the ball for a scrummage.

The teams consisted entirely of amateurs, a stipulation made by the RFU in 1886. However, in the early 1890s, some northern clubs were accused of paying players to compensate them from loss of work.

Many of the teams were set up by local industrialists and players had to leave work early on a Saturday in order to play. A proposal was made by some Yorkshire clubs that players should be allowed compensation, known as 'broken time' payments, but this was rejected by the RFU. Charging for admission to games was also not permitted. It was not until one hundred years later that the RFU were to decide to permit professionalism in the game.

In 1895, twenty-two northern clubs met in Huddersfield to form the Northern Rugby Football Union, which, some years later, was to become the Rugby League. Changes were made to the rules, largely to make the game flow better: reducing the number of players from fifteen to thirteen, heeling the ball back after a tackle rather than stopping for a scrummage, and the elimination of the line-out.

The NRFU also set up a league competition structure, and to differentiate teams playing in the RFU, which did not have a league, and the NRFU, which did, the latter were described as playing in the league, hence the name Rugby League as opposed to Rugby Union.

The game has gained huge international popularity around the world and in 2013 the final of the Rugby League World Cup between Australia and New Zealand was played at Old Trafford in front of a massive 74,468 spectators.

Badminton

Although invented in India, the game we know today as Badminton was created by British military officers stationed there.

The game was developed from an earlier game called battledore (bat) and shuttlecock where, similar in some ways to today's badminton, the object was to strike the shuttlecock with a bat as many time as possible before it hit the ground. This was a non-competitive game often played by groups of children.

The British took this simple game and added a net – although this may have originally been a string stretched between two poles – and also added a court. The game was popular in the Indian town of Poona (now Pune), where many British soldiers were stationed, and seems to have been known initially as Poona. The first set of rules were drawn up there in 1873.

The game was brought to Britain by returning Army officers and in the same year it was introduced by the Duke of Beaufort at a garden party held at his home in Badminton. Thus it was initially known in Britain as the Badminton game; not difficult to see where its name came from.

In those early days the game was played both indoors and outdoors on an hourglass court, but this was later changed to a rectangular court and as a competitive sport it is played indoors.

Servicemen returning from India were keen to continue playing the game, and the first Badminton club appears to have been set up in Folkestone in 1875, still using the original rules drawn up in Poona. However, in 1887 J. H. E. Hart, at the Bath Badminton Club, drew up a set of rules more appropriate to the game being played in Britain. The number of clubs continued to grow and in 1893 fourteen clubs got together to set up the Badminton Association. As with similar associations in other sports, they set about standardising the rules and setting up a competition to find the best players – the All England Badminton Championships – in 1899. The first champion was Sydney Howard Smith, also a top-class tennis player.

As badminton spread to other countries the need arose for an international organisation to govern the sport, and in 1943 the International Badminton Federation was set up in Kent. The organisation became the Badminton World Federation and is now headquartered in Kuala Lumpur.

In 1939 Sir George Thomas, a tennis player who switched to badminton, donated the Thomas Cup to be awarded to the top men's singles champion. Not to be outdone, the Uber Cup for women was created in 1956, named after Betty Uber of England, one of the top doubles players. In 1992 Badminton became an Olympic sport following its demonstration in the 1972 and 1988 Olympics, and it has been included in the Commonwealth Games since 1966.

The federation now has around 150 members, split into five regions throughout the world. Clearly Badminton is firmly established as one of the most popular international sports.

Bowls

The game of bowls, as played on a grass surface, is hundreds of years old. The first bowling green, which is still in use today, is the Southampton (Old) Bowling Green, which dates back to 1299.

Of course, the most famous game of bowls ever was played in Plymouth by Sir Francis Drake. On 18 July 1588, he was playing his favourite game at Plymouth Hoe when a message came through that the Spanish Armada was approaching the coast. Legend has it that he said there was time to finish the game and beat the Spaniards too.

The game became so popular that in 1361 King Edward III banned the game because archery, vital to the nation's defence, was being neglected in favour of bowls. The game was restricted to wealthy people who paid a fee of £100 to have a green and this could be used only for personal and private use. Various subsequent acts confirmed the ban on bowls, including an Act in 1541 that banned the playing of bowls by artificers, apprentices and the like, except at Christmas and then only in their master's house and in his presence. Anyone playing bowls outside their own garden or orchard was liable to a fine of eight shillings and sixpence. This Act was not repealed until 1845, although it was not strenuously enforced.

The game continued to be played by those having the means to do so. It is reported that King James I of England (who was also King James VI of Scotland), as well as King Charles I, Samuel Pepys and others, enjoyed the game and placed bets on the outcome.

However, until the early eighteenth century Scotland was independent and not subject to the bans. As a result the game flourished north of the border. Hundreds of bowling greens were created and the game became hugely popular. Of course

Bowling on the Green, the Bowls Club, Stanley Park in Blackpool. (© VisitEngland/Visit Blackpool)

in the winter months it was not possible to play the game so the Scots, ever resourceful, invented a similar game that could be played using stones sliding over ice – the game we know as curling.

Like other British sports, they spread across the world with the help of the military and the empire. The game was played in America in the seventeenth century and George Washington is said to have been a fan. In Canada the game was first played, appropriately perhaps, in Nova Scotia in 1730, and it reached Tasmania in 1844.

Back in Scotland, in 1864 William Mitchell published his *Manual of Bowls Playing*, which laid down the rules for the game as we know it. As a boy, Mitchell is reported to have played at Scotland's oldest club, Kilmarnock Bowling Green, which dates back to 1740. Today the international federation governing the sport, World Bowls, is still based in Scotland

and has fifty-one member national authorities in forty-six member nations.

And Then There Is Crown Green Bowls …
The game of bowls is traditionally played on a flat green and is played up and down in a number of lanes. However, crown green bowls is played on a lawn which is higher at the centre and the game is played all over the lawn. The game is particularly popular in the north of England and is arguably a more complex game given the nature of the playing surface.

The sport is well organised and inter-county matches date back to 1893. Since 1908 the organisation of the game has been the responsibility of the British Crown Green Bowls Association.

And Indoor Bowls …
Bowls is primarily an outdoor sport played on rain-free days. In order to continue playing in unsuitable weather, the game of Indoor Bowls was devised. Played on mats the length of the normal bowls rinks, the game is largely the same.

And Short-Mat Bowls …
However, it is not always possible to find an indoor venue with sufficient space for full-length mats, so a modified game of short-mat bowls was also devised. The game is again similar to the outdoor game except that there is a bar halfway along the mat that bowlers must go round, using the bias of the ball. This prevents them from bowling a fast straight ball.

The game was created by two South African bowlers who came to work in south Wales. Frustrated by the weather and the lack of playing opportunities, they began using a strip of carpet in a church hall. They subsequently moved to Northern Ireland, where the game became established as a popular sport. It was brought to England by Irish expatriates but was slow to catch on. However, in 1984 the English Short Mat Bowls Association was established, which promoted it as a sport suitable for all ages and requiring little in the way of cost.

> The arrival of the lawnmower, patented in 1830, greatly helped to prepare the grass surface for bowls as it did for tennis and football.

Squash

Squash was developed from the earlier game of rackets, a game played by prisoners who, to pass the time, hit a ball against a wall using a racket.

The game was, perhaps surprisingly, taken up by public schools, including Harrow, who maybe by accident invented a game that was to eclipse rackets and become a game played around the world.

The game of squash gets its name from the soft ball it uses. It dates back to around 1830, when pupils at Harrow School discovered that using a punctured rackets ball, which squashed on impact with a wall, produced a more exciting game, a greater variety of shots and the need to run a lot more as players did not have time to wait for the ball to bounce and come back to them.

Indeed, the modern squash ball comes in grades of softness which affects the speed of the ball and thus the speed of the game. This is denoted by a coloured dot on the ball. The softest and therefore the slowest ball, giving the fastest game, is indicated by a double yellow dot. This is the one used for matches and experienced club players, whereas faster balls with red, blue, white and single yellow dots are used by beginners, intermediate players and children, although sometimes a faster ball will be used on very cold courts as the temperature of the ball significantly affects its ability to bounce.

The game initially proved popular and in 1864 the first squash courts were built at Harrow School. Later, there was even a squash court in the first-class section of the *Titanic*.

The first written mention of squash other than at Harrow School, appears to have been in a book entitled *The Badminton Library of Sports and Pastimes*, written by the Duke of Beaufort and published in 1890.

As to court size, the standard court was proposed as one used at the Bath Club, measuring some 32 feet by 21 feet, considerably smaller than a rackets court, but it was not until 1923 that these dimensions were agreed.

As with all sports there needed to be some form of codification and in the UK this role was carried out by a subgroup of the Tennis and Rackets Association. It wasn't until 1928 that the Squash Rackets Association was formed.

The game spread internationally as a result of British influence around the world, much of it through the armed forces. As a result of this, many top players came from countries where Britain had a strong influence. In 1933, the Egyptian Amr Bey won the first of five British Open championships, then widely regarded as world championships, to be followed by fellow Egyptian Karim who won four British Open championships between 1947 and 1950. However, no feature on squash would be complete without the mention of the Khan family from Pakistan, who produced a series of world champions – Hashim (1951), Roshan (1957), Hashim (1958), Azam (1959), Azam (1962), Mohibullah (1963), Jahangir (1982) Jahangir (1992) Jansher (1993) and again Jansher (1994).

Nevertheless, squash was a niche sport, largely associated

with public schools, until the 1960s when its rapid adoption in the mass market resulted in it becoming hugely popular, nowhere more so than in Britain, in the 1960s to the 1990s. By that time worldwide participation had seen the sport grow to an estimated 46,000 courts being used by 15 million people, this despite the fact that as it is played in a closed court it is not a good spectator sport and not ideal for television coverage. However, this has improved a little in recent years with the introduction of glass courts.

Its attractiveness in part is that it is played indoors, so no concerns about the weather. It is also highly competitive, takes up little space, the equipment is relatively cheap and as one of the most energetic sports ever it enables players to get a great workout in thirty to forty minutes. Post-match activities often involve opponents buying each other a drink and there is also a good social aspect to squash clubs.

In 1998 squash was included in the Commonwealth Games in Kuala Lumpur but as yet it has not become an Olympic sport; perhaps in part due to its limited spectator appeal. The game almost made it in 2012, when London hosted the Olympics, and it has been missed out in the 2016 games in Rio. However, there is a concerted bid by the governing body to get it included in 2020.

Squash has been described as exciting, exhausting and explosive. What more could you want from a sport?

Squash provides an excellent cardiovascular workout. In one hour of squash, a player may expend approximately 600 to 1,000 food calories.

Rounders

As with all ancient sports, nobody is really sure when rounders began. A similar game to that played today has been played in England since Tudor times. The earliest known mention of rounders, along with its descendant baseball, is in a book from 1744 called *A Little Pretty Pocket-Book* by John Newbery, where it was called 'base-ball'.

In 1828 William Clarke published the second edition of *The Boy's Own Book* in London, which included rules of rounders and contained the first printed description in English of a bat and ball base-running game played on a diamond. The following year the book was published in Boston, Massachusetts.

The first nationally formalised rules were drawn up in 1884 by the Gaelic Athletic Association (GAA) in Ireland, where the game is still regulated by the GAA. In the United Kingdom it is regulated by Rounders England, which formed in 1943. While the two associations are distinct, they share similar elements of gameplay and culture.

Competitions are held between teams from both traditions, with games alternating between codes and one version being played in the morning and the other in the afternoon. After the rules of rounders were formalised in Ireland, associations were established in Liverpool and Scotland in 1889.

Rounders is linked to British baseball and is still played in Liverpool, Cardiff and Newport. Although the game is assumed to be older than baseball, literary references to early forms of 'base-ball' in England predate the use of the term 'rounders'. The game is now played up to international level.

The first organisation associated with the sport of rounders was the Liverpool and Scottish Rounders Association, formed in 1889. The National Rounders Association, formed in 1943, was designed to encourage and promote the game. This organisation is also responsible for the rules of the game and for training umpires.

Rounders can be played outdoors, on grass, artificial surfaces, on the beach or indoors in a sports hall. There are several variations ranging from the traditional nine-a-side, two-innings game to fun mini-versions.

Lawn Tennis

Thomas Henry Gem, better known as Harry Gem, was born in May 1819. Educated at King's College London, he began practising as a solicitor in Birmingham in 1841.

A member of the Bath Rackets Club, it is believed that he met Juan Bautista Augurio Perera, a Spanish-born merchant, while playing there. Combining their experiences of rackets and the Basque game pelota, Gem devised a new rackets game which they played on Perera's croquet lawn at his home in Edgbaston in 1865.

In 1872, Gem and Perera moved to Leamington Spa where, together with two doctors, they established the world's first lawn tennis club. Games were played on the lawns of the nearby Manor House Hotel.

However, much of the credit for making tennis the world-famous sport it is today belongs to **Major Walter Clopton Wingfield** who, in 1873, designed and patented a similar game called *sphairistikè*; the word is Greek for 'ball-playing'. Wingfield was a genius at marketing. He produced boxed sets of all the equipment needed to play the game including net, poles, rackets, balls and a set of rules. Using his connections in the aristocracy and legal profession, he sent out a number of sets around the world.

Wimbledon (© VisitLondon/BritainonView/PawelLibera)

In 1874 an American returned from Bermuda with a *sphairistikè* set and, keen on watching British Army officers playing, laid out the first tennis court at the Staten Island Cricket Club. The first American championship was held there in 1880, and the singles title was won by an Englishman named O. E. Woodhouse.

The Wimbledon Championship, the world's oldest tennis championship, was first held in 1877 at what is now the All England Lawn Tennis and Croquet Club – note the continuing link to croquet dating back to that original game on Perera's croquet lawn. The winning champion was Spencer Gore, who was runner-up to Frank Hadow the following year.

Until 1906, all Wimbledon Champions were British and the event was dominated by Lawrence Doherty (five championships) and his brother Reginald Doherty (four championships). The British grasp on the championship was broken by the Australian Norman Brookes in 1907, although Britain staged a comeback in the following two years with Arthur Gore. British players failed to win another championship until Fred Perry won three in a row from 1934 to 1936 and the next British winner was Scot Andy Murray in 2013.

In those early days, tennis also became popular in France and Australia where championships were also held, although until 1925 the French championships were open only to those players who were members of French clubs. Today these four historic championships – Wimbledon, French, US and Australian – are the most important and are known as Majors, or Grand Slams, a term borrowed from the card game bridge.

The rules of tennis, looked after by the International Tennis Federation, have hardly changed since 1924 and this has greatly assisted the development of tennis as a worldwide sport.

Originally only amateur players could participate in championships, but in 1926 a professional tour comprising mostly French and American players was set up and these matches were played to paying audiences. However, it was not until 1968, when championships became 'open', that professional players could participate. This meant that all tennis players could earn a living from the game and this, together with TV rights and the establishment of a new international tennis circuit, has made the game hugely popular to a global audience.

Yellow tennis balls were used at Wimbledon for the first time in 1986.

Modern Olympic Games

William Penny Brookes of Much Wenlock, born 1809, is considered to be the father of the modern Olympic Games.

Brookes, a surgeon, magistrate, botanist and educationalist who was born, lived, worked and died in the Shropshire market town, also became a Justice of the Peace in 1841 and remained an active magistrate for over forty years. His confrontations with cases of petty crime, drunkenness and theft in the local community almost certainly influenced his desire to develop structured physical exercise and education for the working classes and young people. He believed that physical activity not only improved young people's health but that it could also be used to unite and inspire communities.

He devoted his life to improving the lives of the people of Much Wenlock, but it was his idea to revive the Olympian ideal that was to have a worldwide impact and lead to the modern Olympics.

The Olympics in a rare hand-drawn 1948 postcard. (© British Library)

He founded the Wenlock Agricultural Reading Society (WARS) in 1841 for the 'diffusion of useful knowledge', which included a library for working-class subscribers. Interest groups met at the Corn Exchange, which was the WARS headquarters, and in 1850, the Olympian Class was formed to encourage athletic exercises, including everything from running to football, by holding an annual 'games' offering prizes for sports competitions.

Following the 1860 Games, the Olympian Class separated from WARS due to a difference of opinion between the two organisations, and changed its name to Wenlock Olympian Society (WOS) to emphasise that it was now independent.

His lifelong campaign to get physical education on the school curriculum brought him into contact with young French aristocrat Baron Pierre de Coubertin, who visited Much Wenlock in the 1890s to understand how sports education was practised in schools. While there he learned of Dr Brookes's ambitions to host international games to be held in Athens.

The society staged a games especially for the baron and the local people battled for honours in pursuits such as quoit-throwing and cricket. This inspired Coubertain as he began to realise the potential for the modern Olympics. Coubertain was impressed with what he saw and sat up with Brookes long into the night discussing how the Wenlock games might be translated on to a bigger stage.

Coubertin wrote, 'If the Olympic Games that Modern Greece has not yet been able to revive still survives there today, it is due, not to a Greek, but to Dr W. P. Brookes.'

Brookes did not see his dream become reality as he died just four months before Baron de Coubertin helped set up the International Olympic Committee in 1896, which was followed by the Olympic Games in Athens, which came under the auspices of the committee. Much of what happened at the first Modern Olympiad was based on Brookes's ideas. The 1896 Olympics were open to all sportsmen from across the world and featured 280 participants from thirteen nations, competing in forty-three events without restrictions based on social class. However, women were not at this stage welcome to compete in the games.

The concept of moving the Olympics from city to city was also based on another of Brookes's ideas. As well as his town contest, the doctor had helped establish the National Olympian Association (NOA), which featured yearly sporting festivals held across England. One of these was at London's Crystal Palace in 1866, where bearded giant of Victorian sport W. G. Grace took part in front of 10,000 spectators. Each year the NOA games were held in a different town or city, and the place which hosted them financed the meeting, as is the case with the Olympics today.

In 1994, then President of the International Olympic Committee Juan Antonio Samaranch laid a wreath on Brookes's grave, saying, 'I came to pay homage and tribute to Dr Brookes, who really was the founder of the modern Olympic Games.'

Ancient Olympic athletes competed in the nude.

6

Miscellaneous

Christmas Cards

Bath-born **Henry Cole** was a civil servant with a finger in many pies. From 1837 to 1840 he worked as an assistant to **Rowland Hill** – of whom more later – and played a key role in the introduction of the Penny Post. That was when it was decided that a letter should cost the same to deliver anywhere on the mainland. In fact, Henry is also sometimes credited with the design of the world's first adhesive postage stamp, the Penny Black. The name gives a clue as to their colour – black – and cost – an old penny. They had Queen Victoria's head on them in white.

At Christmas in 1843 Henry was too busy to write a personal holiday greeting to his friends and acquaintances and he decided to ask London artist John Calcott Horsley to design a card he could send to everyone.

Henry's cards made use of religious symbolism regarding Christmas. They had three panels with the two outer panels depicting people feeding the poor and clothing the naked. In the centre, a family was eating a huge Christmas dinner. Horsley also painted sprigs of holly, the symbol of chastity, and ivy, symbolic of a place where God has walked, throughout the design. Not that it pleased everyone. It was criticised by temperance groups because the family pictured had wine glasses raised in a toast, but as is often the case with controversy it only served to make card sending more popular.

Henry was interested in industrial design and under the pseudonym Felix Summerly designed items which went into production, including a prize-winning teapot manufactured by Minton. As Felix Summerly, he wrote a series of children's books, including *A Book of Stories from the Home Treasury*; *A Hand-book for the Architecture, Sculpture, Tombs, and Decorations of Westminster Abbey*; *An Alphabet of Quadrupeds*; and *The Most Delectable History of Reynard the Fox*.

In 2001, one of Cole's first Christmas cards, sent to his grandmother in 1843, sold at auction for £22,500.

Henry began his career at the age of fifteen at the Public Record Office, where he became assistant keeper. He was instrumental in reforming the organisation and preservation of the British National Archives.

Christmas Crackers

While on the subject of Christmas, it wasn't only cards that were invented by the British. We were also responsible for Christmas crackers, the brainchild of Londoner **Tom Smith** in 1847. He created crackers as a development of his bon-bon sweets, which he sold in a twist of paper, the origins of the traditional sweet wrapper. As sales of bon-bons dried up, Smith started to come up with new promotional ideas. His first idea for improvement was to insert 'love messages' inside the sweet wrappers. Then one evening when he was sitting by the fire and heard a log crackle, he thought that was something else he could incorporate. With these additions, the size of the wrapper had to be increased and then eventually the bon-bon itself was dropped from the idea, to be replaced by a trinket. The new product was initially marketed as the *Cosaque* but 'cracker' soon became the commonly used name, as rival varieties came on the market. Other elements of the cracker as we know it, that is the gifts, paper hats and plastic toys, were introduced by Walter, Tom's son, to distinguish his product from that of rival cracker manufacturers. Seeing a popular idea, they started to spring up everywhere.

The longest Christmas cracker-pulling chain is 603 participants and was achieved by the RuneScape Community, at Tobacco Dock, London, on 2 November 2013.

Christmas crackers. (© Cgros841 under Creative Commons 2.0)

Modern Postal Service

Rowland Hill from Kidderminster was the founder of our modern postal service. He opened up what had been a complex and expensive system to a much wider public. At that time letters were normally paid for by the recipient not the sender, and of course the recipient could simply refuse to take delivery. Prepaying postage was voluntary and much of the time the postman was left trying to find the addressee in order to redeem the cost. Hill said that costs could be reduced dramatically if the sender prepaid the postage. Many people opposed his scheme, but in 1840 the Penny Black was born and within a few years the number of letters sent through the post had rocketed from 76 million to nearly 400 million.

Sir Rowland Hill statue, Kidderminster. (© Green Lane under Creative Commons 2.0)

The total print run for the Penny Black was 286,700 sheets, with 68,808,000 stamps. A number of these have survived largely because envelopes were not normally used. Letters in the form of letter sheets were folded and sealed with the stamp and the address on the obverse. If the letter was kept, the stamp survived. The only known complete sheets of the Penny Black are owned by the British Postal Museum.

Adhesive Postage Stamp

'A bit of paper just large enough to bear the stamp, and covered at the back with a glutinous wash.' This was how Rowland Hill described the world's first adhesive postage stamp, but although Rowland was determined to go down in history as the inventor of the postage stamp, Scotsman **James Chalmers**, a weaver from Arbroath, is recognised by the *Encylopaedia Britannica* as the true inventor. Chalmers moved to Dundee in 1809 and became a bookseller, newspaper publisher and printer. Later he served as a burgh councillor and became Convener of the Nine Incorporated Trades. As such, he was described as a slayer of the 'dragons which retard progress', battling repeatedly in the cause of burgh reform, and fighting for the repeal of taxes on newspapers and newspaper advertisements.

However, his main interest was postal reform and from 1825 he campaigned for the authorities to speed up the mail between Edinburgh and London by convincing them that this could be done without costing more. In December 1837, he sent a letter outlining his proposals to Robert Wallace, MP for Greenock. He suggested the adhesive postage stamp and cancelling device and submitted his idea to Parliament in 1838. This made Rowland Hill's Penny Postal Service a practical proposition. Chalmers did not favour the use of an envelope for a letter, as each additional sheet incurred an additional charge. Instead, he proposed that a 'slip' or postage stamp could seal a letter.

The inscription on Chalmers gravestone says, 'Originator of the adhesive postage stamp, which saved the penny postage scheme of 1840 from collapse.'

Crossword Puzzle

Liverpudlian **Arthur Wynne** settled for a time in Pittsburgh, Pennsylvania, where he worked on the *Pittsburgh Press* newspaper and played the violin in the Pittsburgh Symphony Orchestra. He later moved to New York City to work on various newspapers, ending up at the *New York World*.

His editor asked Wynne to create a new game for the Sunday 'Fun' section. As a child, Wynne had played a game called 'Magic Squares'. Set in ancient Pompeii, the game's goal was to arrange words that read the same way across and down. Wynne took the basic concept, added a larger, more complex grid and gave the player clues to help solve it. He also pioneered the use of adding blank black squares to the puzzle.

On 21 December 1913, the first crossword puzzle, which was diamond-shaped, appeared in newspapers. Wynne didn't call it a 'crossword' puzzle though; he named it the 'wordcross'. It was only changed to 'crossword' after a typesetting error.

Whatever it was called, it was a huge success with the readers. Soon other newspapers were running the puzzles too. Initially, the only major American daily to refuse to use the puzzle was the *New York Times*, but the crossword finally found its way into the paper's Sunday edition eighteen years after the puzzle's introduction.

Arthur didn't copyright the idea so made no money from it.

The Bank of France

Strange though it seems, it was a murderer, playboy, inveterate gambler and Scot who founded the Bank of France.

Scotland's banking history spans three hundred years and is littered with a series of innovations, including the overdraft, first mutual savings bank, first double-sided banknotes, first banking institute and first professional body of accountants. Another first added to that list was courtesy of Scottish economist and gambler **John Law** when he established the Banque Générale in France in 1716. Law, a brilliant mental calculator who won card games by mentally calculating the odds, was appointed Controller General of Finances of France under King Louis XV.

Born in Edinburgh to a family of bankers and goldsmiths, Law had a sound financial grounding, having been trained since he was fourteen in his father's business. He grew up to be a brilliant man with an incredible mind, but unfortunately he was also a profligate gambler, a womaniser and even a murderer.

In 1688, when his father died, Law left Scotland and high-tailed it to London, where he indulged in his gambling habit and flirted with the ladies. In fact, his eye for the ladies almost got him hanged after he upset a rival for the affections of aristocrat Elizabeth Villiers, who became Countess of Orkney and who was the acknowledged mistress of King William III. On 9 April 1694, he fought a duel over the lady and killed his opponent, Edward 'Beau' Wilson. He was taken to trial and initially found guilty of murder, which in those days carried the death sentence. He was extremely lucky when his sentence was commuted to a fine, on the grounds that the offence only amounted to manslaughter. Wilson's brother appealed and had Law imprisoned, but Law managed to escape to the Continent. He lived something of a nomadic existence, wandering round Europe's gambling houses, though he did also further his financial studies as well as working on a scheme for economic revolution.

He went back to Scotland around 1700 and went to the Scottish Parliament with his proposals. His ideas were rejected because they were too radical for the financial brains of the time. While there he published a book, *Money and Trade considered, with a Proposal for Supplying the Nation with Money* (1705).

In 1715 he came to the attention of the Duke of Orleans, who was regent for the young French king. France was in a dreadful mess, facing its third bankruptcy in a century as a result of expensive wars and King Louis XIV's extravagant lifestyle. Law's ideas, that the French economy would be turned around through credit and the introduction of paper money, appealed to the pressured French duke and in 1716 the Banque Générale was created.

It was a private bank but three-quarters of the capital consisted of government bills and government-accepted notes, effectively making it the first central bank of the nation. Law believed that money was only a means of exchange that did not constitute wealth in itself. He said that national wealth depended on trade, that money creation would stimulate the economy, that paper money was preferable to metallic money which he declared should be banned and that shares are a superior form of money since they pay dividends.

Law's bank proved a great success. The capital was divided into shares with banknotes promising to pay the bearer the value specified on the date of issue. By 1717 the banknotes were accepted as a means of paying taxes, and the following year, when the bank became the Banque Royale, the notes were guaranteed by the king. The French economy was stabilised. Law was hailed as a financial genius.

It seemed as if Law's lucky star was shining, encouraging him to take a risk. This would prove to be his undoing. In 1717 he set up the *Compagnie de la Louisiane ou d'Occident* to exploit the – what seemed to be – limitless resources of the Mississippi basin. His Mississippi Scheme attracted European investors encouraged by his success and the thought of the wealth in the American colony.

Before long the company had overtaken most of its competitors and became the *Compagnie des Indes*. It became so powerful that it merged with the Banque Royale. Law became the French Controller General of Finances. But it was not going to last.

In 1720, after shares soared to 18,000 livres in a frenzy of speculation, the wisest investors cashed in their stock. It became clear that the falsely inflated value of the company and the national bank had outstripped the available capital. The result was a disaster. France went bankrupt overnight. The debacle sent ripples across Europe and was a contributing factor to the start of the French Revolution.

Disgraced and impoverished, after rescuing and ruining the Bank of France, John Law took to the road once more, dying penniless in Venice in 1729. What a sad end for the man with the brilliant mind who created the Bank of France.

Raincoats

On rainy days, what would we do without our macs? We have **Charles MacIntosh**, a chemist from Glasgow, to thank for keeping us dry. In 1822 he was experimenting with ammonia in his efforts to find a way to make rubber pliable. He decided that naphthalene, a by-product of coal tar, was a better and less smelly option than some to soften rubber, so decided that was what he would use. He coated a thin fabric with the sticky solution and sandwiched it between two layers of fabric. Voila! Waterproof material. His family started selling the coats as the 'Macintosh' and that is what they have been known as ever since.

And, do you know that though the technology is now very old, it is still used to make the classic mac product beloved of designers? It takes four years to learn how to create a Macintosh from beginning to end. Garments are hand-stitched, then seams are taped over with thin strips of fabric to ensure they are completely waterproof. Of course, there are imitations and modern macs are made from everything from polyurethane laminates to patent PVC, industrial-strength nylon and tactile microfibre.

Oh, and why does the raincoat sometimes have the addition of a 'k' in the spelling when Charles didn't have a 'k' in his surname? Simply that over the years, writers often – erroneously – added the letter to describe the raincoat and it stuck.

Although the Macintosh style of coat has become generic, a genuine Macintosh coat should be made from rubberised or rubber-laminated material.

Wellington Boots

Of course, if it rains and you wear a mac, you might possibly also put on your Wellington boots. These waterproof boots made of rubber or plastic were named after **Arthur Wellesley, 1st Duke of Wellington.** He came up with the idea as a more wearable alternative to the Hessian cavalry boot which was popular in the eighteenth century. These 'Wellington' boots as we know them today are sometimes often referred to as simply 'wellies'.

Cardigan

While on the subject of clothes, we mustn't forget the cardigan, for which we have **James Brudenell, 7th Earl of Cardigan,** to thank. This brave soldier led the 'Charge of the Light Brigade' at the Battle of Balaclava. He soon became known as a hero, and as he often chose to wear this knitted article of clothing people wanted to follow suit and wear one too. It didn't have a name, but began to be known as a 'cardigan' after Brudenell.

The Dinner Jacket or 'Tuxedo'

Until the late 1880s, formal evening wear for men consisted of a tailcoat jacket that did not button at the front and matching trousers, worn with a dress shirt with winged collar, white waistcoat and white bow tie, hence the short description of the attire as 'white tie'. The dress code was reputedly codified by Beau Brummell, and left very little discretion to the wearer as to its format.

During the 1880s, the then Prince of Wales decided that this formal and strict attire was not really suitable for all occasions and sought something a little simpler.

The prince had for some time been experimenting with more comfortable forms of jacket, including what we now know as a 'smoking jacket'. It was in the style of a lounge suit jacket but made of blue silk, similar to the material used in dressing gowns. However, in 1886 reports began to appear that a new jacket, tailless and modelled more on the style of a military jacket, with silk-lined facing, was being worn with trousers normally associated with a tail jacket. These were being made by Savile Row tailors **Henry Poole & Co**.

Unlike the tailcoat, the new dinner jacket was worn with a black bow tie, hence the short description of this attire as 'black tie'. It is unclear exactly who owned the first dinner jacket, but many stories refer to a Lord Dupplin who was invited to a bachelor dinner by the Prince of Wales aboard his yacht in Cowes. The story goes that Dupplin asked the prince's tailor, Henry Poole, to make him a jacket appropriate for this dinner and the dinner jacket was born.

Despite being unique at this point, it would appear that the prince liked the idea and adopted it as evening wear in preference to the tailcoat, except on very formal occasions.

Clearly a more comfortable attire than the very formal 'white tie' ensemble, the new dinner jacket or dinner suit quickly caught on.

And if you have ever wondered why, in America, the dinner jacket is called a tuxedo, here follows the generally accepted story of why this is.

Around 1888, a visiting wealthy American, Lorillard Griswold, was invited to dine with the prince. Uncertain as to what he should wear, he was directed to the prince's tailor, who made a dinner jacket for him.

A resident of Tuxedo Park, a gated community outside New York, Lorillard Griswold then wore his newfangled jacket to the Tuxedo Park Ball. Although initially the object of curiosity, Americans also came to understand that this new attire offered many advantages over the old 'white tie' outfit and it grew in popularity.

Today, 'white tie' attire is normally reserved for formal state occasions or ballroom dancing.

White tie and black tie are both evening attire, to be worn after 6.00 p.m. when other activities cease and people gather for dinner. However, in the US black tie is sometimes worn during the day at weddings.

Criminal Fingerprinting

Dr Henry Faulds was a Scottish surgeon working in Japan in 1880. His eureka moment came when he realised he had uncovered the secret to catching criminals. The answer, in fact, was at his fingertips.

Born into a working-class Ayrshire family, Faulds had to leave school at thirteen to find work in Glasgow as a clerk to help support his family. At twenty-one he enrolled at the Faculty of Arts at Glasgow University, where he studied Mathematics and Classics. He later studied medicine and after graduation became a medical missionary for the Church of Scotland.

In 1871 he was sent to India where he worked for two years at a hospital in Darjeeling, and in 1873 he was appointed by the United Presbyterian Church of Scotland to establish a medical mission in Japan. He became surgeon superintendent of Tuskiji Hospital in Tokyo, became fluent in Japanese, taught at the local university and was responsible for founding the Tokyo Institute for the Blind.

In the late 1870s he became involved in archaeological digs. He started to notice how impressions left by craftsmen could be found in ancient clay fragments. Examining his own fingertips and those of others, he discovered that the pattern of ridges was unique to each individual.

It was shortly afterwards that his hospital was broken into. Police arrested a staff member, although Faulds believed the man was innocent. He compared fingerprints left behind at the crime scene to those of the suspect and they were indeed different. On the strength of this evidence, the police agreed to release the suspect.

He spent the next few years devising a system of forensic fingerprinting, but Scotland Yard were not interested and turned it down. However, Argentinian police used it to determine that two young boys had been killed by their mother. That was the first time fingerprints had been successfully used in a murder investigation.

Shortly afterwards, Sir William Herschel, a British civil servant working in India, published a letter in *Nature* magazine where he said he had been using fingerprints as a method of signature before Faulds' discovery. This led to a long-standing battle of words between the two men which went on until Faulds died in 1930, understandably frustrated by what he saw as a lack of public recognition for his achievement.

> Faulds helped introduce Dr Joseph Lister's antiseptic methods to Japanese surgeons. In 1875, he co-founded the Rakuzenkai, Japan's first society for the blind, and helped stop the spread of cholera in Japan.

Genetic DNA Fingerprinting

On the morning of 10 September 1984 it dawned on **Alec Jeffreys**, who was working in his laboratory at Leicester University, that there were significant differences between the DNA of different people. Born on 9 January

1950, Jeffreys studied Biochemistry at Oxford University, graduating in 1972 with a first-class honours degree. In 1977 he moved to Leicester University, where he later made his ground-breaking discovery. Perhaps coincidentally, perhaps not, Leicestershire was also the birthplace of Colin Pitchfork, the first person convicted using DNA evidence. He was convicted of two murders and sentenced to life imprisonment in 1988.

Fingerprints were a great means of identifying people but first the prints had to be obtained. DNA, often taken from tiny blood samples, is more readily obtainable from crime scenes. DNA can survive for many years and has also been used to solve cold cases such as the murders of Rachel Nickell and Damilola Taylor, the identity of the Boston Strangler and the proof that a man who drowned in Brazil over twenty-five years ago was Josef Mengele, known as the Nazi Angel of Death.

However DNA is not just used to solve crimes. One of the early uses of the technique was in 1985, when it was used to establish innocence in a dispute about parentage.

Thirteen-year-old Andrew Sarbah had been living with his father in Ghana for some time and was apprehended at Heathrow in 1983 on his way to meet his mother, Christiana, with whom he was intending to live. Officials believed, incorrectly as it later turned out, that his passport was either forged or had been substituted. He was taken into custody and faced deportation.

Christiana's MP managed to get him released to stay with her and their case was supported by Hammersmith Law Centre during their two-year fight. However, officials still doubted that Christiana was his mother; they thought she was his aunt, and wanted to deport him.

Hearing about the then new DNA testing, the centre asked for help from Alec Jeffreys in 1985, having managed to delay the deportation for two years. Tests matched twenty-five bands of DNA with his mother's, proving that Christiana and Andrew were in fact mother and son.

Kaleidoscope

Although **Sir David Brewster** invented the kaleidoscope, this scientist and inventor left a far greater legacy. Born in 1781 in Jedburgh, Scotland, he was a child prodigy and by the age of ten he had built a telescope aided by mentor James Veitch. By the time he was twelve, he was attending Edinburgh University.

When he grew up Brewster became a minister, but realised he did not like speaking in public so opted for a career in science. His main interest was the physics of light and he became renowned for his research of optics. He had constructed several more telescopes and often built his own instruments or improved existing ones to help his experiments.

Brewster wrote *Some Properties of Light*, which was his first paper accepted by the Royal Society, and it was published in 1813. A major scientific distinction for Brewster came through his studies of the polarisation of light by reflection and biaxial crystals, and 'Brewster's Law' still offers a clear way to calculate the angle at which light must strike a material for maximum polarisation, that is the concentrating of light waves on to one plane.

In 1816 Brewster perfected his kaleidoscope invention and patented it in 1817. The device was successful, and people found it fascinating. Unfortunately, due to problems with the patent, manufacturers were able to mass produce kaleidoscopes without Brewster receiving financial rewards.

In 1832, although he was knighted by King William IV,

Brewster stayed as humble as before, continuing to pursue his experiments.

Scottish novelist James Hogg described him: 'He has indeed some minor specialities about him. For example, he holds that soda water is wholesome (*sic*) drink than bottled beer, objects to a body's putting a nipper of spirits in their tea, and maintains that you ought to shave every morning, and wash your feet every night, but who would wish to be severe on the eccentricities of genius?'

The kaleidoscope was so called from the Greek word '*kalos*' for 'beautiful', '*eidos*' meaning 'form' and '*scopos*' which means 'watcher'.

The Pencil

In the 1500s, an enormous deposit of graphite was discovered near Seathwaite in Cumbria. The locals didn't know what it was. After all, chemistry was in its infancy and everyone thought that maybe it was some form of lead. Anyway, they found it was quite useful for marking sheep.

This deposit of pure and solid graphite could be cut into sticks and was given the name plumbago, Latin for 'lead ore'.

Eventually realisation dawned that the value of this graphite was enormous. For a start it could be used to line the moulds for cannonballs, so mines were taken over by the Crown and guarded. When sufficient stores of graphite had been accumulated, the mines were flooded to prevent theft until more was required.

Graphite had to be smuggled out for use in pencils. Because graphite is soft it needs some form of encasement, so graphite sticks were initially wrapped in string or sheepskin for stability. The news of the usefulness of these early pencils spread far

Colour pencils. (© Pencil Museum, Keswick)

and wide, attracting the attention of artists all over the world. Distinctively square, the English pencils continued to be made with sticks cut from natural graphite into the 1860s.

England continued to enjoy a monopoly on the production of pencils until a Frenchman patented a way of using graphite powder mixed with clay which was then baked together.

The hardness of the graphite varies by the ratio of clay to graphite so a pencil marked 'HB' is hard and black, while a pencil marked 'HH' is very hard. Even though it is possible to make synthetic graphite, a good pencil needs natural graphite.

The town of Keswick, near the original findings of block graphite, still manufactures pencils. The factory in which they are made is also the location of the Cumberland pencil museum. It is where you can see the biggest colouring pencil in the world. It was the idea of technical manager Barbara Murray and was completed on 28 May 2001. It is 26 feet in length and weighs 984.1 lbs.

The world's longest coloured pencil. (© Pencil Museum)

The US Navy

Yes, you read that correctly. And it is possibly one of the most incredible British inventions of them all.

Scottish Sailor **John Paul Jones,** born in Kirkcudbright in 1747 as simply John Paul, was the founder of the US naval force. He later emigrated and fought against Britain in the American War of Independence.

When he was twelve years old he entered the British merchant marine and went to sea for the first time as a cabin boy in the brig *Friendship*. At twenty-one he received his first command, the brig *John*. In 1773 in Tobago he killed the leader of his mutinous crew in self-defence and to avoid trial he fled to Virginia. He was considered a fugitive by the British so concealed his identity by adding the surname 'Jones'.

At the start of the American Revolution he entered the Continental Navy, where he was commissioned first lieutenant serving aboard the American flagship *Alfred* and was first to hoist the Grand Union flag on a Continental warship. It wasn't long before he was promoted to captain and given command of the sloop *Providence*. On his first voyage on the *Providence*, he destroyed British fisheries in Nova Scotia and captured sixteen British ships.

On 1 November 1777, he was given command of the *Ranger*. Sailing into Quiberon Bay in France on 14 February

1778, Jones and Admiral La Motte Piquet changed gun salutes. This was the first time that the Stars and Stripes, the flag of the new nation, was officially recognised by a foreign government.

On 23 September 1779, his flagship *Bon Homme Richard* engaged the forty-four-gun Royal Navy frigate HMS *Serapis* in the North Sea off Flambrough Head. Although his own vessel was burning, Jones would not accept the British demand for surrender, replying, 'I have not yet begun to fight.' Over three hours later *Serapis* surrendered and Jones took command. *Bon Homme Richard* sunk the next day and Jones transferred to *Serapis*.

After the American Revolution, Jones served as a rear admiral in the service of Catherine the Great of Russia.

It wasn't only the US Navy that was established by a Brit. Scotsman Admiral Thomas Cochrane from Lanarkshire, who was nicknamed the Sea Wolf in the Napoleonic Wars, was welcomed to Chile in 1818 where he became a citizen of the country, took command of the rebel navy and introduced British customs. Thomas had been dismissed from the Royal Navy in 1814 following a conviction for fraud. While in charge of the Chilean Navy, he also contributed to Peruvian Independence. In 1832, he was pardoned by the Crown and reinstated in the Royal Navy with the rank of Rear Admiral of the Blue.

Seed Drill

Jethro Tull, who was born in Berkshire in 1674, was a bright boy who went to Oxford. However, ill health meant he couldn't pursue a career in law or medicine so he went back home to work in farming with his father.

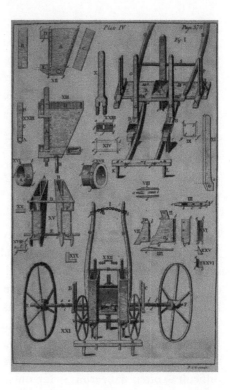

Jethro Tull seed drill,
1731.

Since Roman times, there hadn't really been much in the
way of advancement as far as farming went; labourers still
scattered seed on the land by hand and most seeds were
quickly devoured by birds. Tull knew that things wouldn't
progress unless some radical changes were introduced.

Influenced by the early Age of Enlightenment, he helped
transform agricultural practices by inventing and improving
numerous implements, most important of which was his seed
drill. Pulled by a horse, the drill dug a straight groove into
the soil at the right depth and dropped regularly spaced seeds
into it. It made planting crops more efficient by avoiding
wastage, so hugely increasing the crop yield.

Jethro Tull considered earth to be the sole food of plants. He claimed, 'Too much nitre corrodes a plant, too much water drowns it, too much air dries the roots of it, too much heat burns it; but too much earth a plant can never have, unless it be therein wholly buried: too much earth or too fine can never possibly be given to their roots, for they never receive so much of it as to surfeit the plant.'

Founder of Mitsubishi

Born in Fraserburgh, in the north-east of Scotland, on 6 June 1838, **Thomas Blake Glover** was the fifth son of a family of seven boys and one girl. In Scotland his is not a household name by any means, but it is a different story in his adopted homeland of Japan, where he is seen as a hero.

As a young man he worked with trading company Jardine, Matheson & Co. until he settled in Japan in 1859. At the time the country was viewed as a closed society where business was difficult – often dangerous – for outsiders. Nevertheless, he soon became one of Nagasaki's most influential businessmen and set up his own tea export company in Nagasaki, called the Glover Trading Company.

Before long a more intriguing business opportunity emerged, with rising tensions between the Shogunate and rebellious clans in southern Japan. Glover made a fortune selling ships and arms to rival factions and his arms dealing provided the samurai with the military might to restore the emperor to the throne.

Glover commissioned three warships for the Japanese navy from shipyards in Aberdeen before setting up his own shipyard that grew into the firm of Mitsubishi. He also introduced Japan's first railway locomotive and mechanised coal mine and helped found the Japan Brewing Company, makers of Kirin beer.

When hostilities ended in 1868, Glover, who had been the richest foreigner in Japan, slid into bankruptcy. However, his business interests were diverse enough to allow him to recover and live comfortably for the rest of his life.

Glover married a Japanese girl in 1867. Her name was Tsuru and she is said to have been the inspiration for *Madame Butterfly*, Puccini's famous opera, first performed at the Scala, Milan, in 1904.

When she was seventeen Tsuru had to divorce her first husband, a samurai, due to political differences between her family and his at the time of the overthrow of the Tokugawa Shogunate. She was separated from her baby daughter, Sen. However, there the resemblance between Tsuru and the fictional character ends, for she lived to marry Thomas Glover and they had two children, Hana and Tomisaburo.

The so-called Scottish samurai who helped overthrow the autocratic Shogun rulers of the nineteenth century, Glover was the farsighted industrialist so respected by the Japanese that in 1908 Emperor Meiji awarded him one of Japan's highest honours, the Order of the Rising Sun, Gold and Silver Star. Glover was the first foreigner to be honoured in such a way and he is still remembered there as one of the founding fathers of modern Japan.

Glover House, which was built by Glover in 1863, stands on a hillside overlooking Nagasaki Harbour. The house, which was given UNESCO World Heritage Site status in 2015, is the oldest Western-style building in Japan and nowadays the house and garden, a tourist attraction, are visited by around 2 million Japanese every year. Some years ago, the plaque on the building was changed to describe him as a 'Scotsman' rather than English.

Spectacles

Roger Bacon, a philosopher and educational reformer born in Ilchester, Somerset, was a major medieval proponent of experimental science. He was a clever boy and studied mathematics, astronomy, optics, alchemy and languages. He was the first European to describe in detail the process of making gunpowder. As if that wasn't enough, he also proposed flying machines and motorised ships and carriages.

He is also the one you should thank if you can't read without spectacles. In 1249 Bacon investigated optics and the refraction of light through lenses, leading to the development of spectacles. In those days spectacles did not have rigid arms and had to be tied on with ribbons, or simply held in place – which must have been jolly awkward.

In their infancy, the sight of someone wearing spectacles could cause a sensation. There is a story that when a bespectacled envoy from Padua appeared at the Duchess of Austria's nuptials in 1319, there was so much excitement

Medieval spectacles in the Eglise St Jacques, Rothenburg. (© Daniel71953 under Creative Commons 2.0)

among the guests, the bride was overshadowed. Later spectacles were associated with magic, so jesters and demons were regularly depicted wearing them.

Over the centuries craftsmen developed a wide range of models, and from around 1400 came nose spectacles. These had one-piece wire frames with round lenses. There were also spectacles that could be fixed to caps or hats. All these early aids to vision were usually sold by travelling salesmen and pedlars.

It wasn't until the 1700s that Londoner and optician **Edward Scarlett** made the first glasses with rigid side arms which balanced on top of the ears, the type of spectacles most frequently used today. Demand may have been fuelled to some extent by the availability of cheap daily newspapers. Around 1730 Scarlett advertised that he 'grindeth all manner of Optick Glasses [and] makes spectacles after a new method, marking the Focus of the Glass upon the Frame, it being approv'd of by all the Learned in Opticks as [the] Exactest way of fitting different Eyes'.

An engraving by William Hogarth, 'Characters Who Frequented Button's Coffee House about the Year 1720', illustrates the social background well. It depicts Joseph Addison and Martin Folkes, who is holding nose spectacles in his hand. A parliamentary news sheet headed *Votes of the Commons* lies on the table. The snuff box and tobacco pipe indicate two other types of gentleman's accessory, the development of which to some extent parallel that of spectacles.

There were times when spectacles were considered hideous and bizarre. In the eighteenth century numerous ways were often found to hide them when not in use, in rings, fans or perfume bottles.

At the beginning of the nineteenth century the monocle was introduced, becoming popular in diplomatic circles.

By then special medical glasses were also in use, and later developments included protective spectacles for railway travellers. Spectacles were also used as a stylistic device by artists and by the nineteenth century they were regularly associated with intelligence.

In the 1750s, mover and shaker **James Ayscough**, a medical instruments designer, invented hinged spectacles with tinted lenses in blue and green to reduce glare and treat some vision problems. These spectacles are believed to be the precursors to sunglasses. It seems the eighteenth century was as cool as today. That's not all he made, though; he also made microscopes.

In days gone by, sailors believed that wearing a gold earring as opposed to a pair of glasses would help improve their eyesight. Maybe that is why pirates are often depicted wearing gold earrings.

These days glasses are usually made from plastic rather than glass to prevent cracking and damage to the eyes in the event of an accident.

Matches

In 1826, **John Walker**, a chemist from Stockton-on-Tees, accidentally discovered that a stick coated with chemicals burst into flame when scraped across his hearth at home. He thought it was an idea he could develop.

His early experiments involved coating wooden splints or sticks of cardboard with sulphur tipped with a mixture of sulphide of antimony, potassium chlorate and starch, the sulphur serving to communicate the flame to the wood. From there he went on to invent the first friction match and sold his first 'Friction Light' in April 1827 from his pharmacy in Stockton-on-Tees to a Mr Hixon, a solicitor in the town. Walker's first friction matches were made of cardboard, but before long he began to use wooden splints cut by hand. Later he packaged the matches in a cardboard box containing a piece of sandpaper for striking. There is no denying they were dangerous as balls of flame sometimes fell to the floor, burning carpets and clothes, which led to them being banned in France and Germany. Walker did not patent his 'Congreves', as he called the matches in an allusion to the Congreve's rocket invented in 1808.

But to be truthful Walker wasn't the first to try this, because records tell us that in 1680 Irish physicist Robert Boyle coated a piece of paper with phosphorous and a small piece of wood with sulphur. He rubbed the wood across the paper to create a fire, though he did not pursue his development and did not create a usable match.

Scottish inventor **Sir Isaac Holden** came up with an improved version of Walker's match. The exact date of his discovery, according to his own statement, was October 1829. Holden did not patent his invention, and as a result London chemist **Samuel Jones** copied the idea and launched his own 'Lucifers' in 1829.

The original price of a box of fifty matches was one shilling. A piece of sandpaper folded double was supplied with each box, through which the match had to be drawn to ignite it.

Fire Extinguisher

The first recorded fire extinguisher was patented in England in 1723 by chemist **Ambrose Godfrey**. It was a cask of fire-extinguishing liquid and contained a chamber of gunpowder connected to a fuse system. When ignited, the gunpowder exploded so scattering the solution. It was probably used to a limited extent, as on 7 November 1729 *Bradley's Weekly Messenger* refers to its effectiveness in helping extinguish a London fire.

The modern fire extinguisher was invented in 1818 by **George William Manby** from Norfolk, who was a captain in the Cambridgeshire Militia. His extinguisher consisted of a portable copper cask containing three gallons of potassium carbonate solution contained within compressed air. It was given the name The Extincteur.

Manby also invented what came to be known as the Manby Mortar after witnessing a ship running aground off Great Yarmouth in 1807. This device fired a thin rope from shore into the rigging of any ships in distress. A strong rope when attached to the thin one could be pulled aboard the ship. His invention was adopted in 1814 and mortar stations were established around the British coast. It was estimated that by the time of his death nearly 1,000 people had been rescued from stranded ships by means of his apparatus.

Manby was the first to advocate a national fire brigade. He was elected a Fellow of the Royal Society in 1831 in recognition of his many accomplishments.

Elastic Fabric

Thomas Hancock from Marlborough founded the British rubber industry. In 1815 records show that he was in partnership in London with his brother, Walter, as a coachbuilder. Thomas's interest in rubber appears to have come from the desire to make waterproof fabrics to protect the passengers on his coaches, and by 1819 he had begun to experiment with making rubber solutions.

He went on to patent elastic fastenings for gloves, suspenders, shoes and stockings but in the process of creating these early elastic fabrics realised that he was wasting huge amounts of rubber. This led to his invention, a wooden machine which used a hollow cylinder studded with teeth. Inside the cylinder was a studded core that was hand-cranked. The machine shredded rubber scraps, allowing rubber to be recycled after being formed into blocks or rolled into sheets.

Hancock was very protective of his invention and never patented it. He even referred to it as his 'pickling machine' so that people wouldn't know what it was used for. It is now known as a 'masticator', which sounds a bit rude, but of course it isn't.

An endnote to Hancock's story: you might have thought with all that spare rubber around he would have thought to invent the rubber band, but he didn't. **Stephen Perry** took out the patent for that on 17 March 1845.

Thomas Hancock went on to supply Charles Macintosh with rubber for his raincoats after the advantages of the masticator caught his interest. In 1825 Hancock applied for a licence to use Macintosh's naphtha process and the two men went on to become partners in the manufacture of waterproofed items.

The Whistle

Can you imagine what it must have been like to have to depend on a wooden hand rattle to summon help? It must have been something of a losing battle to say the least, but that's what the Metropolitan Police relied on once upon a time. By 1883, they knew a replacement would have to be found. The rattles they used were weighty, bulky and pretty useless since the noise they made didn't penetrate far in London's bustling, pea-souper-clogged streets. They couldn't think what to do about it, though, so that same year they decided to hold a competition to try and find a better way to attract people's attention.

This is where **Joseph Hudson** comes in. Originally from Derbyshire, he moved to Birmingham, where he trained as a toolmaker. He converted his humble washroom at St Marks Square, which he rented for one shilling and sixpence a week, into a whistle-making workshop and in 1870 founded the company J Hudson & Co., which would later become the worlds largest whistle manufacturer.

Hudson entered his design into the competition. The whistle he had made was small but had a distinctive, loud sound which carried much further than rival whistles.

Joseph had come across the distinctive sound by accident when he dropped his violin one day. As the bridge and strings broke he realised the subsequent noise was the sound he was looking for. Enclosing a pellet inside the whistle created the unique warbling sound, by interfering with the air vibration.

When tests on Clapham Common showed it could be heard from up to a mile away, the police knew this was what they had been looking for. Prior to this, whistles had only been used as musical instruments or as toys.

Hudson called his trading company 'Acme' from the Greek word for 'pinnacle' and went on to design whistles for locomotives and football referees who had up until then had to resort to waving a handkerchief or a stick, or simply shouting.

Joseph, supported by his son, continued to revolutionise the world of whistles and in 1884 he launched the world's first reliable 'pea' whistle, 'the Acme Thunderer', which is still the most popular whistle in the world, used by train guards, dog handlers and police officers. In fact, his whistle is still used by the force and many others worldwide. Over 160 million Thunderers have been manufactured by Hudson & Co., which is still based in Birmingham.

Acme supplied the whistle and megaphone for the *Titanic*.

Spiral Hairpin

You have to admit neither its inventor, **Ernest Godward**, nor the product are names which are on everyone's lips, but Ernest Godward has such an amazing story that it more than deserves inclusion in this book.

Godward, the son of a firefighter, was born in London in 1869. Because he was a sickly child he had very little formal education. He did not start to go to school until he was nine years old and he only stayed for three years before deciding enough was enough and running away to sea. He worked on a cabling ship sailing between Nagasaki and Vladivostok, though his seafaring adventure didn't last long and he was sent home from Japan on the orders of the British Consul.

Returning to London, he started an apprenticeship to train as a mechanic with a firm of hydraulic engineers and fire-engine manufacturers. Three years later the urge to travel hit him again and he decided to set sail, this time as a steward on P&O passenger ships. His work at sea consisted mainly of odd jobs: he polished brass, scrubbed the decks and became the crew's 'napkin-folding champion'.

Who would have imagined that in his lifetime he would invent a whole host of items? Let's hear it for the burglarproof window, mechanical hedge clipper, rubber hair curler, an egg beater which beat eggs faster than anything had ever done before, and, of course, spiral hairpins. As if that wasn't enough, he was a leading world expert on the internal combustion engine, became a champion cyclist, excellent musician, talented artist, runner, swimmer, boxer, an accomplished singer, entrepreneur and adventurer who formed an entertainment group called Star Variety Company. To sum up, Ernest Godward was a brilliant overachiever.

He emigrated to New Zealand in 1901 and that's where he invented the spiral hairpin. This predecessor to the hair clip fixed itself to hair, unlike the then common straight pin, which required constant adjustment otherwise it fell out. The spiral hairpin was an immediate success. Godward set up Godward's Spiral Pin and New Inventions Company. He was now a full-time inventor. He travelled to America to secure patents for his inventions, forming the Merkham Trading and Koy-Lo Hairpin companies to take care of manufacture and distribution. He sold the American rights to the spiral hairpin for £20,000, nothing less than a fortune in 1901.

At the same time, he was also a keen promoter of motor racing. With friend and fellow thrill-seeker Robert Murie, he won the Invercargill to Dunedin motor race in a car fitted with an economiser and in 1907 they also piloted the first hot-air balloon to take off from Invercargill.

London authorities were interested in fitting the city's buses with Godward's Eclipse petrol economiser. This was a forerunner of the modern carburettor for motorcycles, cars, buses, tractors and aeroplanes. He travelled to Britain in 1913 to promote it. His contacts were amazed that a man with no engineering qualifications could invent such a device. However, although a factory was established at Kingston-upon-Thames to manufacture economisers, for some reason it did not take off.

Godward returned home and studied engineering while continuing to refine his invention. He then moved to New York and sold his economiser to the US Army in 1926. It was a great invention and meant that vehicles could run on petrol, gas, fuel oil and even bootleg liquor. Three years later, 580 buses and thousands of taxis were fitted with his economiser.

There was certainly a lot more to Ernest Godward than a hairpin.

> Godward invented seventy-two models of the economiser and by the 1930s was recognised as the world's leading authority on the internal combustion engine.

The Rejuvenator

Otto Christoph Joseph Gerhardt Ludwig Overbeck – no, his name doesn't suggest he was British, but he was born in London, was a British national and lived in Salcombe, Devon. He studied Chemistry at Cambridge and became known as a rather eccentric inventor after he patented the Rejuvenator, a device that uses electrodes to 'practically renew youth'. His bizarre gizmo can be seen at Thackray Medical Museum in Leeds.

He commissioned the prestigious Ediswan Company to manufacture the Rejuvenator on a large scale and took out patents in eleven countries. It was designed so that members of the public could treat their own ailments. Overbeck also gave theatrical public demonstrations of the Rejuvenator, including one in 1930 at the Savoy Hotel in London. The machine worked by massaging the skin with an electric current and it was said that besides giving the user a more a youthful look, it could cure everything except germ-borne diseases.

He worked for a time in the brewing industry, where as well as machines for use in brewing beer he patented an early rival to the beef drink Bovril. In fact, his yeast-based 'Nutritious Extract' was a prototype of Marmite.

Climate Change Theory

James Croll was born in 1821 in Perthshire, Scotland. He taught himself physics and astronomy at Glasgow University library where he worked as a janitor. Twenty-five years later he sent papers to science editors who recognised their quality and they were published.

Since he put the university as his address, they assumed he was a professor. When they found out that he was in fact a janitor, they couldn't believe it. His work was so impressive that he was made a fellow of the Royal Society and he was given a job with the Geological Survey.

Croll was first to suggest that changes in the shape of the Earth's orbit from elliptical to nearly circular to elliptical again might explain the onset and retreat of ice ages. His theory received support in the 1970s when rhythmic shifts in the Earth's angle of orientation to the Sun and changes in orbit were recognised as creating cool summers, which in turn have been identified as key triggers of ice ages.

And Finally ... The Welsh Discovered America (well, it probably deserves a mention)

... In fourteen hundred and ninety-two/ Columbus sailed the ocean blue...

Columbus, famous Italian adventurer, persuaded the Spanish to back an expedition across the Atlantic. Thereafter generations of schoolchildren were taught that Columbus was the first European to discover America. But, was he really?

There is a story which tells us that long before 1492, in 1170 to be precise, Welsh prince Madog ab Owain Gwynedd sailed from Wales with his followers in search of new lands. They were believed to have landed at Mobile Bay, Alabama, and travelled up the Alabama River along which there are several forts said by the local Cherokee Indians to have been constructed by 'white people'. The structures have been dated to several hundred years before Columbus' arrival and, tellingly, are also architecturally similar to the Welsh Dolwyddelan Castle.

So, did Madog, Prince of Snowdon, really find America? Legend? Tall story? Perhaps not ...

An Indian tribe called the Mandans was discovered in Alabama in the eighteenth century. These white men had constructed forts, towns and villages laid out in streets and squares. They claimed ancestry with the Welsh and spoke in a similar language. In 1837 the tribe was wiped out by a smallpox epidemic introduced by traders. A memorial erected at Port Morgan, Mobile Bay, Alabama, reads, 'In memory of Prince Madog, a Welsh explorer, who landed on the shores of Mobile Bay in 1170 and left behind, with the Indians, the Welsh language.'

Mandans fished from coracles, an ancient type of boat still found in Wales today. Also, unlike members of other tribes, they grew white-haired with age. In 1799, Governor John Sevier of Tennessee wrote a report in which he chronicled the discovery of six skeletons encased in brass armour bearing the Welsh coat of arms.

It is an interesting tale, but perhaps no one will ever know the truth.

To complete this section, here are some more great british inventions in the Miscellaneous category:

Plastic

Invented in the early 1860s by prolific inventor **Alexander Parkes**. Although he took out around eighty patents during his career, he is best remembered for his work with plastic. He was not first to develop a plastic substance, but was the first to demonstrate its qualities outside the scientific community. He displayed the material which he called Parkesine at the International Exhibition in London in 1862, and his company laid the foundations of the plastics industry.

Decimal Fraction

Sixteenth-century mathematician **John Napier**'s discovery of the logarithm has brought misery to generations of maths students. That wasn't all he discovered, though. Napier, an aristocrat and 8th Laird of Merchiston, also invented 'Napier's bones', which was an abacus used to calculate products and quotients of numbers.

Calculus

Calculus is the study of change, with the basic focus being on rate of change and accumulation. In Latin, the word means 'small stone', because it is like understanding something by looking at small pieces. Differential calculus cuts something into small pieces to find how it changes. Integral calculus joins the small pieces together to find how much there is. This was developed during the seventeenth century by physicist **Isaac Newton**. Calculus and its basic tools of differentiation and integration serve as the foundation for the larger branch of mathematics known as analysis. The methods of calculus are essential to modern physics and to most other branches of modern science and engineering.

The Colour Purple

Something a little different. In 1856, **William Perkins,** a London chemist, was trying to synthesise quinine, a malaria treatment made from the bark of the cinchona tree. He didn't succeed in his attempts, but when he was doing so, he mixed coal tar extracts with 'aniline', a substance used in making dyes and explosives. The resultant combination was a strong purple colour. Until then, purple dye was so expensive to produce that only royalty and bishops could wear it. Perkins patented his invention and opened a factory in London where he worked with his father. The colour purple made him a rich man.

Stainless Steel

The idea was mulled over in 1821, but stainless steel was not invented until 1913. Credit has to go to Sheffield man **Harry Brearley** after he was approached by a small-arms manufacturer to find a material that could prolong the life of their gun barrels. The story goes that he disposed of some experimental steel in a yard and a few weeks later found it still as shiny as new. He looked into the reason and discovered that by adding chromium to molten iron, a metal that didn't rust would be produced. Stainless steel is now seen everywhere from surgical instruments and cutlery to turbine blades and architectural cladding.

Sign Language

Scottish schoolteacher **George Dalgarno** was the one who invented the first sign language alphabet. He published two books; one, from 1661, was on a universal language *(Ars Signorum – the Art of Signs)* and the second, nineteen years later, covered methods for teaching the deaf *(Didascalocophus: Or, the deaf and dumb man's tutor)*. Dalgarno advocated writing as a natural method for teaching the deaf, believing that written language for the deaf could be developed as a language for children with normal hearing. He placed great emphasis on early intervention, and advocated finger spelling of words in space through the use of a glove with letters inscribed on it.

Cement

Invented in 1824 by **Joseph Aspdin,** a bricklayer from Leeds. He patented a method of making what he called Portland cement, which is the kind most widely used today. The process involved burning limestone, mixing it with clay and burning it again; the burning produced a much stronger cement than just mixing together limestone and clay.

Piano Foot Pedals

East Lothian Carpenter **John Broadwood** is credited with developing the foot-pedal method for sustaining the piano's sounded notes. He is founder of Broadwood & Sons, the oldest and one of the most prestigious piano companies in the world. Broadwood also revolutionised the instrument's boxy design, coming up with the grand piano in 1777. He must have been a fit young man too, because he walked from his home to London, a distance of almost 400 miles, to work for the harpsichord maker Burkat Shudi. When Shudi died Broadwood took control of the company in 1783. Broadwood's other technical innovations in piano manufacture included adding a separate bridge for the bass notes. He patented the piano pedal in 1783.

Steinway grand piano pedals. (© fanoftheworld under Creative Commons 2.0)

Paraffin

A natural petroleum seepage from the sandstone roof of a Derbyshire coal mine led to Glaswegian chemist **James Young's** discovery of paraffin in 1847. Young found that by slow distillation he could obtain a number of useful liquids. He named one 'paraffine oil' because at low temperatures it congealed into a substance resembling paraffin wax. This meant that homes without electricity could be lit and heated. Young's discovery was a godsend to millions of people. Young also made significant discoveries in rust proofing ships in 1872, which were later adopted by the Royal Navy.

Shorthand

Sir Isaac Pitman from Wiltshire devised the dots, dashes and strokes which we know as shorthand. There are other variations of shorthand, including Gregg, Teeline and speedwriting, but Pitman's is the most widely used. He wrote a letter to his brother in 1837 to say that he had 'contrived a very good alphabet ... a scientific work which I think will stand the test of years'. He was right. His system classifies speech sounds into groups. Consonants are shown as strokes and vowels as dots and dashes. Pitman, a teacher, taught at a private school he founded in Wotton-under-Edge. Through his own publishing house he published manuals, journals and books about shorthand. His motto was 'Time saved is life gained'. He was also vice-president of the Vegetarian Society.

Discovery of Planet Uranus

In 1781, **Sir William Herschel,** who built his own telescopes, discovered a new planet from his home in Slough. This was the first planet to be discovered in modern times and was later named Uranus. It is the seventh planet from the Sun, twice as far out as Saturn. Its mean distance from the Sun is 1,783 million miles. Occasionally, Uranus can become bright enough to be seen with the naked eye, though you have to know exactly where to look. Herschel noticed from the motions of double stars that they are held together by gravitation and that they revolve around a common centre, thus confirming the universal nature of Newton's theory of gravitation. He understood that the whole solar system is moving through space and was able to indicate the point toward which he believed it to be moving. As well as Uranus, he also discovered the lesser-known Mimas, Enceladus and Titania.

Rawlplugs

Named after the guy whose idea they were, **John Joseph Rawlings**. Occasionally they are known as wall plugs. He patented and trademarked them by 1913. By 1919 he had established a company, Rawlings Brothers, to manufacture them but the name was changed to Rawlplug Ltd.

John Bull

This national personification of Great Britain, in particular England, was the creation of **Dr John Arbuthnot**, a Scot from Kincardineshire. Bull, a symbol of Englishness, first appeared as a character in a series of political satires which Arbuthnot wrote. John Bull is depicted as a rotund gent, wearing a tailcoat, breeches and a Union Flag waistcoat.

Modern-Day Inventions

This, the last section in the book, takes a look at what kind of ideas and gadgets today's inventors are coming up with. From the weird and wonderful to the fun and fanciful, British ideas continue unabated. Granted, these inventions may not be in the same category as radar, chloroform, the telephone or the World Wide Web, but they are creative, inspired and undoubtedly fill gaps in certain areas of the market. What's more, the ladies are finally getting a look in!

Sock-Ons

Kezi Levin, a graphic designer and mother of six small boys, has taken her ingenious and practical product '**Sock-Ons**' to market. Kezi studied at Central Saint Martin's College of Art and Design.

Looking after her family, Kezi learned how to be time-efficient. She says, 'I am constantly thinking of how to make life easier. I have always been a mad inventor and my mind spins around at 100 miles per hour.'

It was when her babies wouldn't keep socks on their feet for more than a minute at a time that desperation reached new levels. 'After extensive research I had found nothing to deal with the lost sock/cold feet syndrome and knew I had to come up with a solution. The Sock-Ons idea came while I was trying to fall asleep that night: I got up, cut up

Sock-Ons. (© Kezi Levin)

my own tights, made a prototype and it worked.' Levin's product of softly woven elasticated material that 'locks' babies' socks gently and firmly in place was designed out of sheer frustration.

She was advised to sell the invention to a producer and walk away but decided against that route. 'The one thing that kept me going was hearing that I was crazy to pursue this so far and the more I heard that, the more I felt I was going to do it.'

Over 20,000 Sock-Ons have been sold to date. Levin's message is, if anyone says you can't, say you can. 'I was told that I would never finish my degree, or do so many other things, because of the commitments I was juggling. Take it as a challenge and make the most out of every situation.'

... and still on the subject of babies, here is the Shnuggle Baby Bath ...

Winner of the Red Dot design award, a prize awarded by a panel of design experts, and the Baby Products Association BANTA award, the Shnuggle bath is at the cutting edge of design with its iconic curves, ergonomic design and attention to detail.

As parents, Shnuggle founders **Adam and Sinead Murphy** knew what a baby bath should do: keep the baby warm, safe and upright. The bath features a large foam backrest for comfort, the clever 'Bum Bump' for support and rubber feet for grip. Adam said, 'We spent a lot of time ensuring that the bath met our strict design criteria – modern, innovative and practical.' They created a soft foam material that insulates the water. The bath is textured inside to ensure the baby is stable, and outside to allow for easy grip.

Shnuggle bath. (© Adam Murphy)

The Shnuggle bath has been designed with baby and parent in mind and is suitable from birth up to twelve months. Giving the baby support when they need it and when the baby is ready, it gives the confidence to sit forward and play. The bath is lightweight and easy to carry even while filled with water.

What's more, it comes in a range of colours so that it stands out from traditional plastic white baby baths.

Anywayup Cup

Mandy Haberman specialises in baby-feeding accessories. Originally trained as a graphic designer at London's St Martin's School of Art, she underwent a career change when her youngest daughter was born in 1980 with Stickler's Syndrome and had severe difficulty feeding. Mandy improvised a way to feed her, developing the idea further and patenting the Haberman Feeder for babies with feeding problems. She set up a mail-order business and supplied hospitals and parents direct.

Then, when a visiting toddler ran across a carpet with a conventional children's trainer cup in hand, leaving a trail of pink stains, Mandy realised a cup that sealed by itself when it came out of the child's mouth was needed. So she developed the Anywayup cup technology, set up a company (this time with sales support) and brought the cup to market in 1996.

To be viable, high-volume sales were needed and that meant getting into supermarkets. Mandy was so confident in the cup's performance that she and her team went out on a limb. They put an Anywayup cup filled with concentrated blackcurrant juice in a white box, unprotected by waterproof wrapping, and mailed it to the buyer of Tesco. They included a note: 'If this reaches you as a soggy mess, we've shot ourselves in the foot, but if it reaches you without spilling, please give us a call.' The ploy worked, and the cup was in stores within weeks.

Approximately 50 million cups using Mandy's technology are sold each year globally. Anywayup has won numerous awards for design and innovation.

Mandy was Female Inventor of the Year 2000 and a finalist in the Veuve Clicquot Business Woman of the Year Awards 2003. She was recognised by HM Queen Elizabeth as a Pioneer to the Life of the Nation and has an honorary doctorate from the University of Bournemouth.

The Chronophage Series

Dr John C. Taylor OBE FREng, born in Buxton, Derbyshire, in 1936, is one of the UK's most prolific inventors alive today, with over four hundred patents to his name. After a successful career in which he invented the cordless kettle, an invention that is used over a billion times each day, Dr Taylor turned his attentions to the creation of unique clocks. The Chronophage series, the mechanism of which is protected by six patents and tells the time in a completely novel way, is one of the world's most unique clock families.

Having spent six years living in Canada during his childhood, he returned home towards the end of the Second World War and attended King William School on the Isle of Man before studying Natural Sciences at Cambridge University. After finishing his education he took a job at Otter Controls, founded by his father, and began working in bi-metal. His work with these controls led to Dr Taylor designing the thermostat systems that are used in almost 2 billion kettles and small household appliances.

Mandy Haberman with products. (© Mandy Haberman)

Dr Taylor and the Dragon Chronophage. (© Dr Taylor)

The iconic Dragon Chronophage clock was exhibited early in 2015 at Design Shanghai, China's premier design event, which attracts the global elite of the art, design and architecture sectors. The third in a series of four momentous timepieces, the Dragon Chronophage is a homage to Chinese design. Valued in excess of £3 million, it stands three metres tall and its metalwork is produced to a jewellery standard. Designed in tandem with Professor Long of China, Academy of Art, Hangzhou, the Dragon Chronophage is a rare collaboration between a Chinese designer and a British engineer. This piece of dynamic art slows, speeds up and stops but is always completely accurate every five minutes. It also engages with its audience by rippling its spines, winking and showing off a pearl each hour.

The Dragon is a working component called a grasshopper escapement, which vastly improved the accuracy of the clock when it was invented by John Harrison in the eighteenth century (see section on Marine Chronometer). The grasshopper, usually the size of a coin, has been altered so that it pulls the escape wheel around the circumference of the clock face. This ingenious reworking of a traditional clock mechanism took years to perfect and is protected by six patents.

Taylor studied at Corpus Christi College, Cambridge, and donated significant funds to the construction of a new undergraduate library, the Taylor Library, which was opened in 2008. Wanting to also give something more personal, he began work on the Corpus Chronophage. Featuring a fearsome grasshopper, it was unveiled in 2008 by physicist Professor Steven Hawking. It is installed in an exterior wall of the college and has become one of the city's landmarks.

The Midsummer Chronophage, depicting a huge bluebottle-like insect, was launched at Masterpiece London before being exhibited at Edinburgh's National Museum of Scotland and London's Science Museum and Saatchi Gallery.

The fourth Chronophage is a private commission and details are currently secret at the customer's request.

Each of Dr Taylor's creations to date have celebrated John Harrison's grasshopper escapement through an iconic and mythical creature atop the clock face. Dr Taylor's vision is for future Chronophage clocks to continue this theme, and he is open to new ideas and ways to make this a reality.

Although the three public Chronophage clocks have been plated in 24-carat gold, Dr Taylor is open to working with other materials, such as rhodium, which he used for the graticule of the Corpus Chronophage. A major concern is ensuring that the materials that they use will stand the test of time. Rhodium, a naturally inert metal, will never tarnish, making it an ideal material for Dr Taylor's creations. Dr Taylor acknowledges the dark humour that underpins his creations: 'Each Chronophage, which is Ancient Greek for "time-eater", is relentlessly hungry, gobbling up minute after minute so that you can never have it back. I want each Chronophage to continue to eat time for hundreds of years, so I have used materials that I know will last. Enamel, for instance, has been found in the tombs of the Pharaohs and it is my hope that these clocks will be as mysterious and mythical to our descendants as the hieroglyphs are to us.' By commissioning a Chronophage, a passionate art collector would involve themselves in a unique artistic project, tying their legacy to that of Dr Taylor. In the arts, the commissioners of great works are often remembered alongside the artists themselves.

Dr Taylor has been the recipient of many honours including the appointment as an Officer of the Order of the British Empire in the 2011 New Year Honours list for his services to business and horology, as well as being made a fellow of the Royal Academy of Engineering for his outstanding contribution to the advancement of British engineering, innovation and commerce and being awarded an honorary doctorate from UMIST. In his spare time, Dr Taylor also has a keen interest in aviation, having been a private pilot for over sixty years. Taught by his father as a child, he has so far amassed over 5,000 hours of flying time.

The Alarming Clock

This is the creation of Scottish designers **Natalie Duckett** and **Lee Murray**. Designed to improved sleep rhythm, the clock has two alarms, one signalling in the morning and one in the evening.

The evening alarm signals to the user when to begin their sleep routine to improve the quality and duration of sleep to at least eight hours. When the morning alarm is set, the evening alarm is automatically set to signal nine hours ahead to act as a reminder to consider winding down.

Another interesting feature is that the clock interface is hidden and is on the underside of the product. The clock is made from natural materials to reflect that sleep is a simple and natural process. Each clock is made and assembled by hand on a made to order basis. The touch-sensitive technology simplifies the design further.

A striking feature is its move away from an electronic buzz. The alarm signal is mechanically created to imitate the sound of a woodpecker drumming against a tree. Placing the 'beak' against different surfaces or objects alters the sound, making the alarm signal unique to every home. If an alarm signal is not needed, the 'beak' can be removed.

TravelTime

This is a technology that consumer-facing websites use to let their customers rank and sort data by minutes rather than miles. It is used on various property sites as well as on

VisitBritain. Its creator, **Charlie Davies,** born in Salisbury in 1986, was destined for a colourful career after programming his first BBC computer. He had realised the best learning curve came from breaking something first and then putting it back together, costing his father several hundred pounds when using his father's latest computer equipment as the trial subject. At thirteen, Charlie started his first business by cycling around the New Forest fixing PCs, which led to him building his own technical consultancy business.

When it came to studying, Charlie's inquisitive nature meant that he was always looking to explain how things worked, whether it was people or computers. He decided to get the best of both worlds by studying Politics and Philosophy at Cardiff University, building technical skills on the side.

After graduating Charlie expanded his technology skills. He knew that most location search results are displayed using distance, i.e. 'within X miles', but when people are travelling, minutes mean more than miles. Minutes are not open to interpretation, a mile on the other hand can take one minute or one hour depending on geography, infrastructure or mode of transport. He joined forces with Peter Lilley because he knew two inventor heads are always better than one. Charlie built the TravelTime Platform, the technology that supplies location search sites results by minutes rather than miles, which was granted a patent in 2013. Once the initial idea was in place Charlie set to work making the best out of his invention by expanding modes of transport and geographies. Later the technology fell into the hands of those in the GIS world who realised the software was perfect for aiding businesses in location-based decision making. This resulted in MinuteMapr being born, and questions that used to take days could now be taken in minutes. The TravelTime Platform continues to be a huge success across the search industry, being used in sites including property, travel, dating and recruitment.

The Stylfile Range

Tom Pellereau, an inventor and entrepreneur, studied Mechanical Engineering at the University of Bath, receiving a first-class degree, and a masters in Mechanical Engineering with Innovation and Design. He was winner of the 2011 series of BBC's *The Apprentice*.

While appearing on *The Apprentice*, Tom lost a record eight tasks; regrettably, he says, his ignored ideas could have saved the team had his louder teammates listened. However, as the tasks progressed his inventive ideas shone through.

During the final boardroom meeting, Tom's story of how he first gained a listing in Boots by turning up at the store's headquarters pretending to be a courier and delivering his invention direct to the head buyer gained Lord Sugar's attention. He was crowned the first apprentice to gain investment of £250,000, which probably goes to prove that sometimes the nice guy does win.

Since beginning a partnership with Lord Sugar, Tom has launched his innovative Stylfile range nationwide. The revolutionary best-selling range of beauty accessories includes the Nipper Clipper, which can safely cut the nails of wriggly babies, and the Mani Pro, a nail buffer with stainless steel cuticle pusher to tidy cuticles hidden inside. Then there is the Stylfile Infuse, hailed as the world's first moisturising nail file. This nail file dispenses almond oil to nourish cuticles and nails as you file. The pocket-sized Emergency Fil is a mini-file in a protective, extra-long spiral casing designed to prevent the file from scratching delicate items. It has a strong abrasive suitable for fixing broken nails and a soft 500 abrasive for smoothing nails. Tom made sure that feet

Tom and Lord Sugar. (© Tom Pellereau)

are not left out of the equation either and created the S-Ped. This file's unique curved shape follows the feet's natural contours to give the user smooth soles and heels.

He has also set his creative mind to additional projects, including Mode Diagnostics, which is a screening tool for bowel cancer; Babisil, a fully collapsible silicon baby bottle; and Salvox, a game-changing non-toxic sanitiser capable of neutralising contagious pathogens.

Tom loves to find out what problems people have in everyday life so he can try to solve them to make everyday life easier. He has harboured a lifelong ambition to become an inventor and has a keen passion for design and innovation, something that he credits to time in his childhood spent taking things apart and putting them back together.

He credits his dyslexic mind as the source of his creative thinking, innovative thought process and drive. He struggled with English and languages at school, but luckily his first school recognised his dyslexia early and provided extra lessons.

Tom learnt early that hard work and concentration were vital to fulfilling his potential. Lord Sugar and Tom have now been in business for over three years, and the company they formed — Aventom — is growing.

Inventor Tom has many more ideas in the pipeline.

Bibliography & Sources

Books

Baker, W. J., *Sports in the Western World* (USA: University of Illinois Press, 1988)

Beale, C., *Born out of Wenlock, William Penny Brookes and the British origins of the modern Olympics* (Derby: DB Publishing, 2011)

Burnett, A., *Invented In Scotland* (Scotland: Birlinn, 2012)

Callow, S., *Dickens' Christmas* (UK: Frances Lincoln, 2009)

Claybourne, A. & A. Larkum, *The Story of Inventions* (UK: Osborne Publishing, 2007)

Downey, J. & J. C. Downey, *Better Badminton for All* (UK: Pelham Books, 1982)

Edwards, F., *Cats Eyes* (UK: Blackwell Oxford, 1972)

Evans, E., *British History* (UK: Parragon Books, 2002)

Guillain, J.-Y., *Badminton: An Illustrated History* (UK: Publibook, 2004)

Hannas, L., *The English Jigsaw Puzzle 1760 to 1890, London* (UK: Wayland Publishers, 1972)

Kimpton, P., *Tom Smith's Christmas crackers: an illustrated history* (UK: Tempus, 2005)

Latorre, D. R., Kenelly, J. W., Reed, I. B. & S. Biggers, *Calculus Concepts: An Applied Approach to the Mathematics of Change* (Cengage Learning, 2007)

Lazar, R. & L. Swerling, *How Nearly Everything Was Invented by the Brainwaves* (UK: Dorling Kindersley, 2008)

Leach, J., *From Lads to Lords – A History of Cricket, Volume 1 (to 1914)* (UK: George Allen & Unwin, n.d.)

Lowndes, William., *The Royal Crescent in Bath* (UK: Redcliffe Press, 1981)

Martin, J., 'Shanks, Alexander (1801–1845)', *Oxford Dictionary of National Biography* (UK: Oxford University Press, 2004)

McGinn, C., *The Ultimate Guide To Being Scottish* (Edinburgh: Luath, 2008)

O'Hara, C. B. & W. A. Nash, *The Bloody Mary: A Connoisseur's Guide to the World's Most Complex Cocktail* (Globe Pequot, 1999)

Pope, S., *So That's Why They Call It Great Britain* (UK: Monday Books, 2009)

Russell, L. S., *A Heritage of Light: Lamps and Lighting in the Early Canadian Home* (Canada: University of Toronto Press, 2003)

The London Museum, *Two Hundred Years of Jigsaw Puzzles* (UK: Grosvenor Press, 1968)

Walker, S., *Ernest R. Godward, Inventor* (NZ: River Press, 2013)

Woolley, B., *The Bride of Science: Romance, Reason, and Byron's Daughter* (AU: Pan Macmillan, 1999)

Websites

blog.nationalholidays.com
http://blog.onlineclock.net/chronometer/
http://starleybikes.com/history
http://twyfordshistory.blogspot.co.uk

http://www.10-facts-about.com

http://www.atminventor.com/goodfellow_atminventor.html

http://www.bbc.co.uk/food/recipes/black_bun_66755

http://www.bbc.co.uk/history/historic_figures/bazalgette_joseph.shtml

http://www.bbc.co.uk/shropshire/features/2004/08/william_penny_brookes.shtml

http://www.chemheritage.org/discover/online-resources/

http://www.firstworldwar.com/weaponry/tanks.htm

http://www.flyingmachines.org/cayl.html

http://www.gdfra.org.au/history_of_the_whistle.htm

http://www.hdtrust.co.uk/hist02.htm

http://www.historic-uk.com

http://www.fortnumandmason.com

http://www.historybyzim.com

http://www.jamesbraidsociety.com/jamesbraid.htm

http://www.maclaren.uk

http://www.nzedge.com/ernest-godward/

http://www.oldlawnmowerclub.co.uk/mowers/moms/mp025-shanks-pony-mower

http://www.rampantscotland.com

http://www.rmg.co.uk/explore/astronomy-and-time/time-facts/harrison

http://www.rolls-roycemotorcars.com/history

http://www.sciencemuseum.org.uk/broughttolife/people/johncharnley.aspx

http://www.scotch-eggs.com/news/the-history-of-scotch-eggs

http://www.sirbarneswallis.com

http://www.windsorscottish.com

www.acsu.buffalo.edu

www.Anaglypta.co.uk

www.angusahead.com

www.atminvengtor.com

www.bbc.co.uk

www.bfwbadminton.org
www.brewstersociety.com/brewster_bio.html
www.Britevents.com
www.britsattheirbest.com
www.canadacricket.com
www.college-optometrists.org
www.cruiseholidays.com
www.directoryofbikes.com/bicycle_history.htm
www.dnai.org
www.electricscotland.com
www.encyclopaedia.com
www.engineeringhalloffame.org/profile-young.html
www.examiner.com
www.fifa.com
www.fillyourplate.org
www.foodtimeline.org
www.habsburger.net
www.historic-uk.com
www.Ideafinder.com/history
www.interestedwomen.com
www.japantimes.co.jp
www.jarsandbottles-store.co.uk/blog/company-news/
   the-history-of-marmalade/
www.lawnbowling-arcadia.com
www.Libraryofbirmingham.com
www.Lords.org
www.nationalholidays.com/blog/?p=158
www.navy.mil/navydata/traditions/html/jpjones.html
www.noeperez.net/design/pedaling-history/phbm/hw.html
www.oldlawnmowerclub.co.uk
www.onlinelibrary.wiley.com
www.parksandgardens.org
www.pennyfarthingclub.com/
www.percyshawcatseyes.com

www.pocruises.com

www.pocruises.com.au

www.prnewswire.com

http://waterbedsandfutons.com/the-history-of-the-waterbed/

www.bbc.co.uk/history/historic_figures/bazalgette_joseph.
  shtml

www.radiotimes.com/news

www.rampantscotland.com

www.rolls-roycemotorcars.com/history/

www.squash2020.com

www.rugbyfootballhistory.com

www.SOBG.co.uk

www.sportsknowhow.com

www.telegraph.co.uk/foodanddrink

www.thefa.com

www.tradgames.org.uk

www.undiscovered Scotland.com

www.wanderlust.co.uk

www.worldbowlsltd.co.uk

www.worldsquash.org

www.winningsquash.com

www.Darts501.com

www.wpbf.co.uk

www.writingtimes.co.uk

www.yatesandsuddell.co.uk

www.amazon.com/Written-Blood-Mike-Silverman-ebook/dp/
  B00EKOBXU8

https://forensicbupe.wordpress.com

www.barnesandnoble.com/w/written-in-blood-mike silverma
  n/1113756132?ean=9780552169318

Articles

Bellis, M., 'Joseph Priestley – Soda Water'

Cross, D., 'On the Beat in Birmingham – Rules and Regulations' (BBC, 2011)

Gomez P. F. & J. A. Morcuende, 'Early attempts at hip arthroplasty–1700s to 1950s' (2005)

Gomme, A. B., 'Traditional Games of England, Scotland and Ireland'

Keogh, B., 'The Secret Sauce: a History of Lea & Perrins' (1997)

Maxwell, J. C., 'Experiments on Colour, as Perceived by the Eye, with Remarks on Colour-Blindness', *Transactions of the Royal Society of Edinburgh* (1855)

Norgate, M. & J. Norgate, 'Old Cumbria Gazetteer, Black Lead Mine' (Geography Department, Portsmouth University, 2008)

Arnott, N., 'On the Smokeless Fire-place, Chimney-valves, and Other Means, Old and New of obtaining Healthful warmth and ventilation' (London: Longman, Green, Longman & Roberts, 1855)

Riedel, S., 'Edward Jenner and the history of smallpox and vaccination' (Baylor University Medical Center, 2005)

Williams, G., Lush, P. & D. Hinchliffe, 'Rugby's Berlin Wall – League and Union from 1895 to Today' (London League Publications, 2005)

Taylor, J., 'The Victorian Police Rattle Mystery', *The Constabulary* (2003)

Twyford Bathrooms, 'The Development of the Flushing Toilet – Detailed Chronology, 1596 onwards' (1996)

BBC News, 'The UK's oddest days out' (15 September 2013)

Toole, B. A., 'Poetical Science', *The Byron Journal* (1987)

Waghorn, H. T., 'Cricket Scores, Notes, etc.' (1730–73)

Waugh, W. & J. Charnley, 'The Man and the Hip' (Springer, 1990)

Wilson, N. & A. Murphy, *Lonely Planet Guide to Scotland* (Lonely Planet, 2008)

Woodcroft, B., 'Reference index of patents of invention, from 1617 to 1852' (1855)

Wroblewski, B. M., 'Professor Sir John Charnley (1911–1982)' (2002)

Index of Inventions

adhesive postage
   stamp 193
afternoon tea 29–30
Alarming Clock, the 238
Anaglypta 42–3
antiseptic surgery 153–4
Anywayup cup 233–4
Arbroath smokies 33–4
aspirin 141–2
ATM 130–32
automatic kettle 25–6

baby buggy 45
badminton 174–5
bagless vacuum
   cleaner 51–3
Bank of France, the 195–7
bathroom fittings 10
Belisha Beacon 66–7
black bun 34–5
blood transfusion 160
bouncing bomb 86–9
bowls 176–9

calculus 224–5
carbonated water 54
cardigan 199

catseyes 63–6
cement 226
chequebook and printing on
   metal 17–18
chloroform 137–41
chocolate 17–18
Christmas cards 189–90
Christmas crackers 190–91
chronograph 132–3
Chronophage series,
   the 234–8
climate change theory 222
computer
   programming 108–110
corkscrew 40–41
cricket 162–5
criminal
   fingerprinting 201–2
crossword puzzle 194–5
cruise holidays 92–7
cycle 67–73

darts 166–7
decimal fractions 224
Dewar flask 57
dinner jacket 199–201
disc brakes 104

elastic fabric   216–17
electric light   118–19
electric motor   101
enamel bathtub   58

fax machine   122–3
fire extinguisher   215–6
first practical water
   closet   11
flush toilet   9–10
football   167–72

General Motors Corporation
   of America   74–6
genetic DNA
   fingerprinting   202–4
glider   84–5
golf   161–2
Gray's Anatomy   160

hip replacement   148–9
hovercraft   103
HP Sauce   35–6
hypnotism   146–7
hypodermic syringe   159

ibuprofen   143–4
identifying the mosquito
   as the carrier of
   malaria   157–9
iPod   136

jet engine   85–6
jet propulsion   103
jigsaw puzzle   55–6
John Bull   229

kaleidoscope   204–5

Kendal mint cake   56–7

lawn tennis   183–5
lawnmower   59–62
linoleum   43–4
liquorice allsorts   22

marine chronometer   120–
   21
marmalade   23–4
matches   214–15
microchip   135
military tank   89–90
Mitsubishi   210–11
modern Olympic
   Games   186–8
modern postal service   192
motor car   78
MRI scanner   160

package holidays   97–101
paraffin   228
pencils   205–6
penicillin   145–6
photography   124
piano foot pedals   226–7
piccalilli   57
Pimm's   40
plastic   224
pneumatic tyres   73
portable defibrillator   156–
   7
printing press   128
programmable
   computer   107–8
propeller-driven
   steamship   91
public toilets   12–13

purple   225

radar   121–22
radio   135
raincoats   198
Rawlplugs   229
reflecting telescope   133
refrigeration   48–9
Rejuvenator, the   221
Rolls-Royce   77–9
rounders   182
rugby football   172–3

safety bicycle   71–2
sandwich   27–8
s-bend   11
Scotch eggs   31–3
seaside rock   20–22
seed drill   208–9
sewage system   13
Shnuggle baby bath   232–3
shorthand   228
sign language   226
single-piece ceramic
   toilet   12
smallest mass-market
   transistor radio/pocket
   calculator   129
smallpox vaccine   150–52
Sock-Ons   230–31
sparkling wine/
   bubbly   37–9
spectacles   212–14
speedometer   102
spiral hairpin   219–21
squash   179–81
stainless steel   225–6
steam engine   78–83

steam hammer   134–5
Steri-Spray   154–6
Stylfile Range, the   240–42

Tarmac   74
Teasmade   26–7
telephone   105–6
television   114–18
text messages   125–6
tin cans   37
toaster   24–5
toothbrush   14–16
torpedo   103–4
traffic lights   112–13
typewriter   133–4

Uranus   228
US Navy   207–8

valve actuator   18

waterbed   46–7
Wellington boots   199
whistle   217–18
wind-up radio   135
Worcestershire Sauce   18–20
World Wide Web   110–12